DESIGN IN NATURE

"A most stimulating thought principle, framed in a nice and lively personal story. What I find most exciting is the exceptionally broad perspective that Bejan adopts for developing his concepts. *Design in Nature* is a fascinating read."

—EWALD WEIBEL, Professor Emeritus of Anatomy, University of Bern

"With wide-ranging examples and the iconic pictures to go with them, Bejan illustrates that nature is inherently an outstanding designer of flow configurations, which raises philosophic issues beyond the remit of thermodynamics. Is the distinction between animate and inanimate blurred by their common constructal design? This and many more issues are raised by Professor Bejan's distinguished and original work, fittingly presented in *Design in Nature*."

—JEFFREY LEWINS, Deputy Praelector, Magdalene College, Cambridge University

DESIGN IN NATURE

HOW THE **CONSTRUCTAL LAW** GOVERNS EVOLUTION IN BIOLOGY, PHYSICS, TECHNOLOGY, AND SOCIAL ORGANIZATION

Professor Adrian Bejan *and* J. Peder Zane

DOUBLEDAY

New York London Toronto

Sydney Auckland

 Published in the United States by Doubleday, a division of Random House, Inc., New York, and in Canada by Random House of Canada Limited, Toronto.

www.doubleday.com

Grateful acknowledgment is made to the following for permission to reprint previously published material: page 2: image courtesy of Professor Ewald Weibel, Department of Anatomy, University of Bern; page 120: photograph courtesy of Prof. R. P. Behringer, Department of Physics, Duke University; page 188: photo courtesy of Ron Sherman/Photographer's Choice; page 193: drawing used by permission of the Istituto Geografico de Agostini, Novara, Italy; all other images courtesy of the author.

Jacket design by Michael J. Windsor
Jacket illustration © Daryl Balfour/Gallo Images/Getty Images

Library of Congress Cataloging-in-Publication Data
Bejan, Adrian.
Design in nature : how the constructal law governs evolution in biology, physics, technology, and social organization / Adrian Bejan, J. Peder Zane.—1st ed.
p. cm.
Includes bibliographical references and index.
1. Pattern formation (Physical sciences) I. Zane, J. Peder. II. Title.
Q172.5.C45B45 2011
500—dc23 2011015398

ISBN 978-0-385-53461-1

PRINTED IN THE UNITED STATES OF AMERICA

1 3 5 7 9 10 8 6 4 2

First Edition

CONTENTS

INTRODUCTION

This book is about design in nature as a scientific discipline, centered on a physics law of design and evolution: the constructal law. This law sweeps the entire mosaic of nature from inanimate rivers to animate designs, such as vascular tissues, locomotion, and social organization.

Discovering a unifying law of design in nature was not on my to-do list when I traveled to Nancy, France, in late September 1995. I was a forty-seven-year-old professor of mechanical engineering at Duke University who had come to deliver a lecture at an international conference on thermodynamics. Giving you a sense of how steeped my career was in mechanical engineering, I remember that I had brought flyers announcing the publication of my seventh book, *Entropy Generation Minimization.*

My work took a fateful turn during the prebanquet speech delivered by the Belgian Nobel laureate Ilya Prigogine. Echoing the scientific community's conventional wisdom, this famous man asserted that the tree-shaped structures that abound in nature—including river basins and deltas, the air passages in our lungs, and lightning bolts—were *aléatoires* (the result of throwing the dice). That is, there is nothing underlying their similar design. It's just a cosmic coincidence (Figure 1).

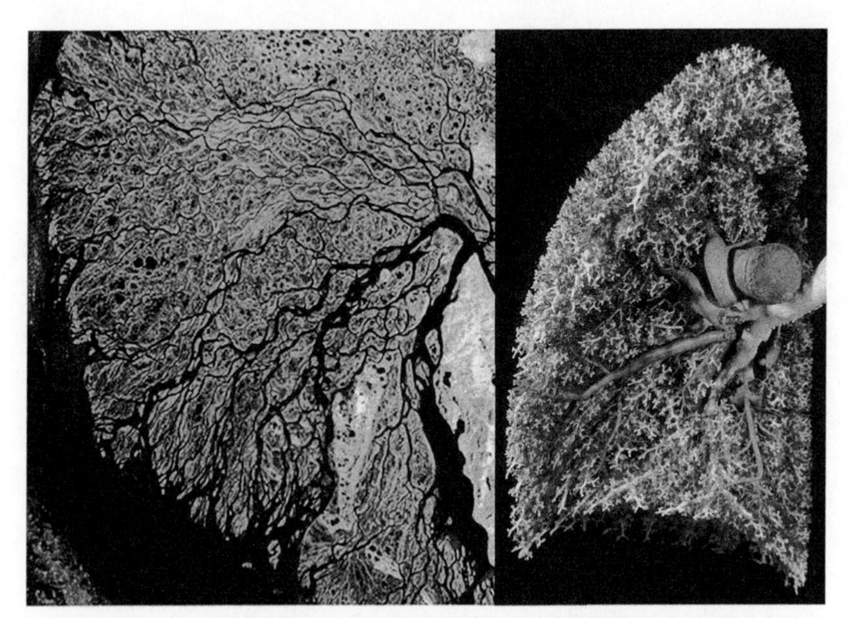

Figure 1. The phenomenon of design in nature unites the inanimate with the animate. The left side shows the delta of the Lena River in northern Siberia. The right side shows a cast of the human lung.

When he made that statement, something clicked, the penny dropped. I knew that Prigogine, and everyone else, was wrong. They weren't blind; the similarities among these treelike structures are clear to the naked eye. What they couldn't see was the scientific principle that governs the design of these diverse phenomena. In a flash, I realized that the world was not formed by random accidents, chance, and fate but that behind the dizzying diversity is a seamless stream of predictable patterns.

As these thoughts began to flow, I started down a long, uncharted, and wondrously exciting path that would allow me to see the world in a new, and better, light. In the sixteen years since, I have shown how a single law of physics shapes the design of all around us. This insight would lead me to challenge many articles of faith held by my scientific colleagues, including the bedrock beliefs that biological creatures like you and me are governed by different principles from the inanimate world of winds and rivers and the engineered world of airplanes, ships, and automobiles. Over time, I would develop a new understanding of evolutionary phenomena and the oneness of nature that would reveal how design emerges without an intelligent designer. I would also offer a new theory for the history of Earth and what it means to be alive.

In addition, I and a growing number of scientists around the world would begin finding new ways to make life easier: better ways to design roads and transport systems; to predict the evolution of civilization and science, of cities, universities, sports, and the global use of energy. We would unravel the mysteries of Egypt's Pyramids and the genius of the Eiffel Tower while demonstrating how governments are designed like river basins and how businesses are as interdependent as the trees on the forest floor.

All that lay in the future when I boarded the plane for the trip home. High over the Atlantic, I opened my notebook (the old-fashioned kind, with paper) and wrote down the *constructal law:*

> For a finite-size flow system to persist in time (to live), its configuration must evolve in such a way that provides easier access to the currents that flow through it.

I was writing in the language of science, but the fundamental idea is this: Everything that moves, whether animate or inanimate, is a flow system. All flow systems generate shape and structure in time in order to facilitate this movement across a landscape filled with resistance (for example, friction). The designs we see in nature are not the result of chance. They arise naturally, spontaneously, because they enhance access to flow in time.

Flow systems have two basic features (properties). There is the current that is flowing (for example, fluid, heat, mass, or information) and the design through which it flows. A lightning bolt, for example, is a flow system for discharging electricity from a cloud. In a flash it creates a brilliant branched structure because this is a very efficient way to move a current (electricity) from a volume (the cloud) to a point (the church steeple or another cloud). A river basin's evolution produces a similar architecture because it, too, is moving a current (water) from an area (the plain) to a point (the river mouth). We also find a treelike structure in the air passages of lungs (a flow system for oxygen), in the capillaries (a flow system for blood), and the dendrites of neurons in our brains (a flow system for electrical signals and images). This treelike pattern emerges throughout nature because it is an effective design for facilitating point-to-area and area-to-point flows. Indeed, wherever you find such flows, you find a treelike structure.

Since human beings are part of nature and governed by its laws, the point-to-area and area-to-point flows we construct also tend to have treelike structures. These include the transportation routes we follow to work (a flow system for moving people and goods), which include many smaller driveways and neighborhood paths flowing into a few larger roads and highways. So, too, do the flowing networks of information, material, employees, and customers that keep those businesses afloat. The engineered world we have built so that we can move more easily does not copy any part of the natural design; it is a manifestation of it. That said, once we know the principle, we can use it to improve our designs.

Although treelike structures are a very common design in nature, they are only one manifestation of the constructal law. In a simple example, logs floating on a lake or icebergs at sea orient themselves perpendicular to the wind in order to facilitate the transfer of motion from the moving air body to the water body. A more complex example is the design of animals that have evolved to move mass better and better (to cover more distance per unit of useful energy) across the landscape. This includes the seemingly "characteristic" sizes of organs, the shape of bones, the rhythm of breathing lungs and beating hearts, of undulating tails, running legs, and flapping wings. All these designs have arisen—and work together—to allow animals, like raindrops in a river basin, to move more easily across a landscape (Figure 2).

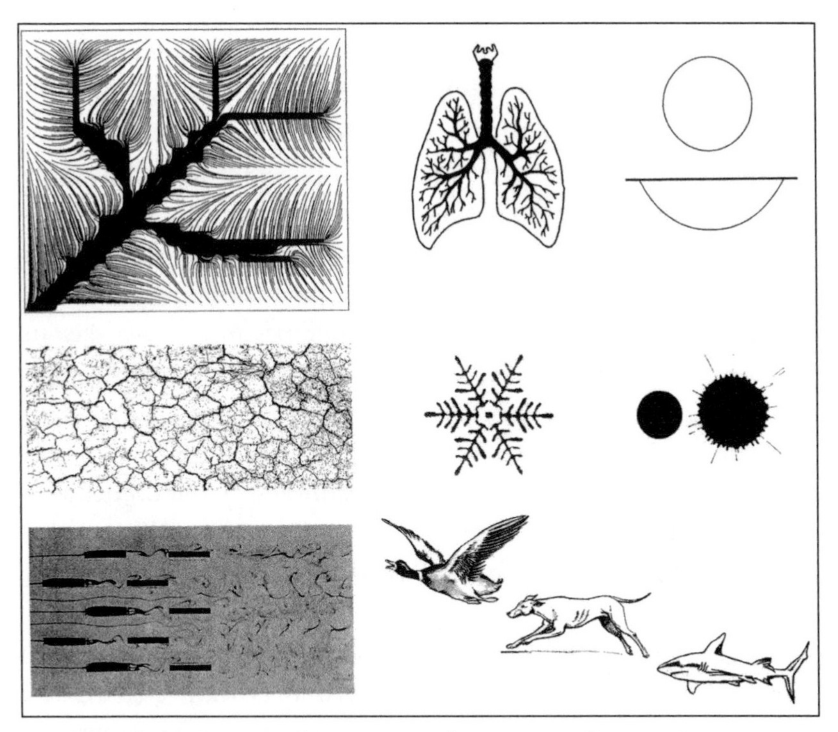

Figure 2. Animate and inanimate phenomena of generation-of-flow configuration in nature, which have been predicted based on the constructal law. *Top row:* river drainage basins, bronchial trees, and round-duct and open-channel cross sections. *Middle row:* cracks in shrinking solids, snowflake solidification, and splat versus splash when a liquid droplet hits a wall. *Bottom row:* laminar versus turbulent flow and animal locomotion (flying, running, and swimming).

The constructal law dictates that flow systems should evolve over time, acquiring better and better configurations to provide more access for the currents that flow through them. Design generation and evolution are macroscopic physics phenomena that arise naturally to provide better and better flow access to the currents that run through them. The majesty of this principle is that it occurs at every scale. Each component of an evolving flow system—each rivulet, each tree, each road—acquires evolving designs to facilitate flow access. As these elements coalesce into larger and larger structures (into evolving river basins, forests, and transport networks), the various-sized components work together so that everything flows more easily. We see this, for example, in the shape and structure of the neural networks in the brain, of the alveoli in the lung, and the human settlements on a map. In the big picture, all the mating and morphing flows on the largest system that surrounds us, the Earth itself, evolve to enhance global flow. *E pluribus unum* (one out of many).

The constructal law is revolutionary because it is a law of physics—and not just of biology, hydrology, geology, geophysics, or engineering. It governs any system, any time, anywhere, encompassing inanimate (rivers and lightning bolts), animate (trees, animals), and engineered (technology) phenomena, as well as the evolving flows of social constructs such as knowledge, language, and culture. All designs arise and evolve according to the same law.

This law tears down the walls that have separated the disciplines of science by providing a new understanding of what it means to be alive. Life is movement and the constant morphing of the design of this movement. To be alive is to keep on flowing and morphing. When a system stops flowing and morphing, it is dead. Thus, river basins configure and reconfigure themselves to persist in time. When they stop flowing and morphing they become dry riverbeds, that is, the fossilized remains of earlier "live" flow systems. The solid, treelike veins of ore found underground today, for example, are fossils of the fluid streams, eddies, and meanders that flowed before solidification a long time ago. Biological creatures are alive until all their flows (blood, oxygen, locomotion, and so on) stop, after which they, too, become fossilized remains.

This unifying definition marks an advance because it removes the concept of life from the specialized domain of biology. It aligns it (or, better, it juxtaposes it) with the physics concept of the dead state, which means "equilibrium with the environment" in thermodynamics: a system that is at the same pressure, the same temperature, and so forth as its surroundings, and hence, in which nothing moves. The constructal law defines life in physics terms, and it covers all live-system phenomena. It also reframes the view that life on Earth began with the rise of primitive species some 3.5 billion years ago. As we will see, "life" began much earlier, when the first inanimate systems, such as currents of solar heat and wind, acquired evolving designs. In the big history of life on Earth, the emergence and evolution of inanimate, animate, and technological designs tell a single story. Where Darwin showed the links between biological creatures, the constructal law connects everything on the planet.

On one level, the constructal law can be expressed through the language of mathematics, physics, and engineering. My colleagues and I have published hundreds of articles in leading peer-reviewed journals. My own books for specialists—including *Advanced Engineering Thermodynamics*; *Shape and Structure, from Engineering to Nature*; and *Design with Constructal Theory*—use the constructal law to predict the phenomenon of design configuration. Leading universities, from Paris and Lausanne to Shanghai and Pretoria, have hosted international conferences and courses on the constructal law.

You don't need advanced mathematics to grasp it. The constructal law is also a way of seeing. Since discovering the law, I have witnessed thousands of people—from renowned scholars and professional scientists to my students at Duke and those at high schools I've visited—experience a moment of discovery like the one I had in Nancy. They, too, hear the penny drop. They see it. They get it. Through this book I hope to help you recognize how the constructal law is shaping everything around—and within—you.

Seeing constructally can be thought of as a three-step process. Step one starts with Leonardo da Vinci's insight that "motion is the cause of every life." I love this quote because it is so expansive. And yet, Leonardo didn't take it far enough, because he was talking only about biological creatures. In fact, not only animals but also rivers, weather patterns, snowflakes, corporations, nations, science, knowledge, culture—you name it—throb and pulse with movement.

Even things that seem just to sit there are, in fact, flow systems. Take that quintessence of stagnation, the mud puddle. There it sits, murky and soupy. And yet, when the sun emerges after the rain, dry air begins to draw moisture from it because of the natural tendency toward equilibrium (in this case, of wet and dry). Before long, the puddle is gone. Soon the dirt begins to crack in telltale, treelike patterns in order to facilitate the flow of moisture from the ground to the air. That puddle is, in fact, a vibrant, morphing flow system. If we trained a movie camera on it, we'd see plenty of action (Figure 3).

Figure 3. Mud cracks on the banks of the Luangwa River, Zambia.

Human beings are also flow systems, similar to but more complex than mud cracks. Internally, the flow of blood carries oxygen and food through a treelike network of blood vessels to organs whose size and shape are just right to enable us to move efficiently per amount of useful energy derived from food. The design of our bodies—just like that of every other animal from sharks to antelopes to great blue herons as well as that of trucks on the highway—has evolved to enable us to cover greater distances per unit of useful energy (food, fuel). And, like trees in the forest, we are also part of other, much larger, flow systems on Earth. When we get in cars, we enter the flow of traffic. In the office, the work we produce flows along with that of coworkers to reach customers through various channels. At the supermarket, tea that flowed from farmers and distributors in Sri Lanka settles into our shopping baskets. As we will see, all these seemingly independent designs are morphing and mating to facilitate our movement.

Step two is to recognize that all flow systems have the tendency to endow themselves with a characteristic that was not recognized until the constructal law—design. This property includes the flow system's configuration (the architecture, geometry, shape, and structure) and its rhythm (the predictable rate at which it pulses and moves).

Design does not emerge willy-nilly. To know why things look the way they do, first recognize *what* flows through them and then think of what shape and structure should emerge to facilitate that flow. The configuration of a flow system is not a peripheral feature. It is the defining characteristic. In later chapters we will illustrate this by showing how the shape and structure of seemingly disparate phenomena—including rivers, fish, sprinters, economies, universities, and the Internet—are predicted by the constructal law.

Step three turns our drawing into a movie because designs evolve. Flow systems configure and reconfigure themselves over time. This evolution occurs in one direction: Flow designs get measurably better, moving more easily and farther if possible. Of course, there will be bumps and mistakes: Every trial involves error. But in broad terms, tomorrow's system should flow better than today's.

This is the natural phenomenon covered by the constructal

law: the generation, ceaseless morphing, and improvement of flow design. This mental viewing enables us to recognize that people, birds, and other animals are flow systems that carry mass on the surface of the globe; that trees and mud cracks are flow systems for moving water from the ground to the air; that universities, newspapers, and books are flow systems for spreading knowledge across the globe. All generate designs that should evolve to better facilitate the flow of these currents. This insight allows us to recognize pattern in phenomena long dismissed as accident.

Consider the snowflake. The prevailing view in science is that the intricate crystals formed by the snowflake have no function. This is wrong. In fact, the snowflake is a flow design for dispersing the heat—called the latent heat of solidification—generated on its surfaces during freezing. As water vapor condenses and freezes it throws off its excess heat. When the ice crystal first forms, its spherical bead is the shape that grows faster than other shapes, the shape that facilitates rapid solidification. When the bead is large enough, needles emerge and enhance solidification (that is, produce ice) faster than the sphere. To facilitate solidification even more, larger snowflakes morph into shapes with more needles that disperse heat. Complexity is finite (modest), and is part of the constructal design that emerges. Complexity is a result, not an objective; not an artist's wish; and, contrary to current dogma based in fractal geometry, it is certainly not "maximized."

Now let's take a closer look at the organized fury of an erupting volcano—a flow system of lava. As it begins its journey through the shaft, the concentration of the mixture of molten rock is such that lava organizes itself into a series of concentric sheaths. In the center is lava of high viscosity (less runny); on the outside is lava of low viscosity (runnier). The low-viscosity lava that touches the solid rock helps it flow. When lava pours out of the volcano, another remarkable phenomenon occurs: The lava seems to select between two flow options, choosing the better way to move at any given time. If the molten rock is moving slowly, it oozes out of the volcano. If it is moving quickly, it generates a different flow configuration—a treelike structure with channels and branches—because this is the better way to move quickly. And, if we know the size of the area that the lava will spread across, we can predict the number of channels that will be generated.

What we are seeing is the mindless lava self-organizing into flow patterns to ease movement. This process happens everywhere in nature. Depending on its size and speed, a falling drop of liquid, for example, will become a splat (round disk) or a splash (crown shape). Smaller and slower droplets come to rest as splats. Larger and faster droplets come to rest as splashes. This phenomenon is well established. Your ink-jet printer, for example, depends on it, emitting specific quantities of ink at just the right speed in order to produce precise images. So does the forensic science of blood splatter popularized through TV crime shows. Before the constructal law, no one knew why this splat versus splash happens. As we will explore later in this book, these two shape-generating ways of flowing—slow and short, fast and long—are ubiquitous. In fact, most systems, including every beat of your heart, every breath you take, and the circuits that power your computer and brain, involve both types of flows. Striking the balance between them is a hallmark of natural design.

The constructal law also teaches us that evolution can be observed at all timescales, including during our own lifetime. When we speak of rivers and animals evolving to increase flow access, we are describing very gradual changes. But when lava generates design, droplets of liquid splash and splat, lightning bolts crackle in the summer heat, and snowflakes form against the winter sky, we are witnessing evolution right before our eyes. We can also watch it occur at home. For instance, if you throw some rigatoni into a pot of boiling water, you can watch the tubes tumble around in a disorganized fashion. After a few minutes, something amazing happens. Instead of lying flat, they begin to stand up straight, organizing themselves into a chimneylike pattern to facilitate the flow of heat and steam. If you

Figure 4. Rice volcanoes: the regular pattern of vertical ducts constructed by the flow of steam during the boiling of rice.

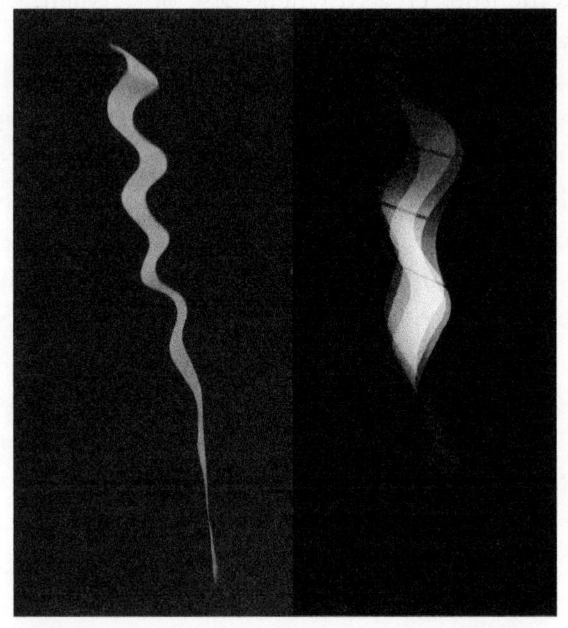

Figure 5. The free fall of a piece of toilet paper makes visible the constructal design phenomenon of turbulence. When the fall is fast enough, eddies of air are configured on both sides of the paper, because this is the more efficient way of transferring vertical motion (momentum) from the paper to the surrounding air. The momentum is transferred laterally, away from the falling airstream. The paper is highly flexible and makes the turbulent eddies visible, looping around them like a skier through slalom gates.

prefer rice to pasta, boil some of that. When the water level drops enough, you will see equally spaced chimneys of steam escaping the entire mushy body. An exquisite tapestry of little volcanoes with round shafts is the easiest way for the heat to come out of the boiling mass, and they form every time (Figure 4). In both cases, the riddle of design is solved by asking what is flowing. The answer is not rigatoni or rice but heat and steam.

Similarly, if you drop a piece of toilet paper from the top of a tall ladder, it undulates so that it falls like a meandering river (Figure 5). Or when you pour a glass of dark beer, regularly spaced eddies emerge around the rim (Figure 6). In both cases, it is not the toilet paper or beer that is generating design but the momentum created when these objects fall. Because of the natural tendency toward equilibrium, the momentum (the movement) is transferred laterally to the surrounding still air and water through the design phenomenon of turbulence. In all instances, design emerges because things flow better with configuration.

Of course, there is no conscious intelligence behind these

Figure 6. More constructal design of turbulence in a glass of dark beer. The momentum from the falling liquid is transferred more effectively to the body of stationary liquid by a design of eddies regularly spaced around the rim. The bubbles gather on the surface only above the regions that correspond to downward flow.

patterns, no Divine Architect churning out brilliant blueprints. To preempt any confusion, let me make this perfectly clear: The constructal law is not headed toward a creationist argument, and in no way does it support the claims of those who promulgate the fantasy of intelligent design. Anyone who takes excerpts from this book to suggest that I am arguing for a spiritual sense of "designedness" is engaging in an intentional act of dishonesty.

Instead, just as other impersonal, naturally arising phenomena such as gravity, the freezing points of fluids, and thermodynamics make things operate in a certain way, flow systems generate better and better flowing designs. Until now, we could only observe the patterns. The constructal law tells us why those patterns arise and empowers us to predict how they should change in the future. It reveals that it is not love or money that makes the world go round but flow and design.

This raises the question: How come? What causes the constructal law? The short answer: We don't know. The constructal law is what is known in science as a first principle, an idea that cannot be deduced or derived from other laws (if it could, it would be a theorem). It just is—a law of physics that governs the emergence of macroscopic shape and structure in nature. Like *all* scientific laws, it is a concise summary that encompasses billions of observations of natural phenomena of the same kind. It addresses two of the biggest questions in science: *Why* does "designedness" (configuration, rhythm, scaling rules) happen *everywhere* in animate and inanimate systems alike? Why does the design-generation phenomenon persist in time?

The constructal law is a shout from the rooftops: Everything that flows and moves generates designs that evolve *to survive (to live)*. This is not a desire or objective but the natural tendency, that is, the physics phenomenon.

As a first principle, the constructal law does not start from observation. It is a pure theory, a purely mental viewing of how things *should be*. We don't catalog and measure every river (or bird, tree, lightning bolt, etc.). Instead, we discover mentally just one of them, and one is enough—it is the cat out of the bag; it keeps us awake until we assure ourselves that nature is the way in which the principle painted it for us in the mind, in the dark of the night. In its streamlined form, our use of the scientific method has three steps:

1. We use the constructal law to predict what should occur in nature—that designs should emerge and evolve in time to facilitate flow access.
2. Armed only with pencil and paper and without any recourse to empiricism (that is, without looking out the window), we determine (anticipate) the right design for whatever is flowing.
3. Later, we go out into the world and compare our predictions to what we find in nature.

To appreciate an important advantage of this theory, consider the work of Robert Elmer Horton (1875–1945), the soil scientist whose achievements were so great that the Horton Medal, the highest honor bestowed by the American Geophysical Union, was named in his honor. One of his achievements was the study of the number of tributary streams that feed each larger river channel. He and his associates spent years poring over empirical data, studying maps, and counting river channels to find that the average number of daughter streams flowing into the mother stream is a number between 3 and 5.

Three colleagues and I found the same scaling rule with pencil and paper using the constructal law. We imagined a very simple river basin and asked what flow structure (in this case, how many tributaries) posed less and less resistance for a given volume rate of water input (the streams) to the territory (river basin). The answer we arrived at was four. No doubt, Horton's empirical work made it easy for us to verify our findings. But had he known about the constructal law, he would not have had to perform innumerable measurements to reach the same conclusion.

Indeed, once we recognize that the constructal law governs design in nature, we can predict all configurations using only our minds. Such is the power of theory.

In the sixteen years since the conference in Nancy, I and many other researchers have not found a single flow system that cannot be predicted by the constructal law. Specialists are using it to illuminate a wide range of subjects, including linguistics and sociology, nuclear decontamination, globalization, finance, warfare, patterns of residential segregation, and human mortality. The applications are so numerous that the constructal law is still in its infancy. You, dear reader, are in on the cutting edge of an emerging idea that has only just started to flow on the globe and into books.

If I were to add two words to the constructal law, they would be these: "given freedom." Constraints abound in our world, preventing things from organizing themselves in more efficient ways. A dam, for instance, stops the river from flowing; bad ideas make it harder for human beings to thrive. I learned that lesson growing up in Romania during the 1950s and 1960s, when it was ruled by a Soviet-imposed government. The Russians had a crummy system and decided to force it on their free and more advanced neighbors. Like all territories, Romania is a flow system for many things, including commerce and ideas. For decades, the communist government choked off those flows, and my native country foundered. The popular uprising in what was then Czechoslovakia led to the Prague Spring in 1968, during which some restrictions were loosened. Romania held a mathematics competition: The six winners from across the nation would be allowed to apply to study abroad. I earned a top score and later was accepted by MIT, where I went on to earn all my engineering degrees. That small access to freedom enabled me—I, too, am a flow system on the landscape—to remake myself, that is, to redesign my movement on Earth.

Rigid governments lacking the ability to change are just one manifestation of the inevitable forms of resistance that obstruct flow. Instead of struggling under dictators or totalitarian governments, flow configurations evolve in one direction in time: to reduce the effects of friction and other brakes that inhibit their flow. Resistance is inevitable and unavoidable. It is why the world will never be a perfect place and why the most flow systems can accomplish is to keep getting better, that is, to be less and less imperfect. Thus the constructal law suggests the idea of progress, conveys the promise of hope: Given freedom, flow systems will generate better and better configurations to flow more easily.

In my academic life, I was particularly attuned to this phenomenon—able to see what others had missed—because I had, quite by accident, grappled with the same problem faced by rivers and trees through my research as an engineering professor at Duke University and as a consultant for industry and government. We engineers are rarely thought of as cool, but my specialty is designing smaller, more efficient systems for cooling electronics. In general, the more computational power you generate, the more heat you create. Run your hand along the bottom of your laptop or the screen of your plasma TV—you could almost fry an egg on them! For decades I used mathematics and the laws of physics to develop better designs for guiding that heat through and out of the box.

I noticed but did not think much of the fact that the drawings I was producing corresponded to the treelike flow structures that appear in nature. Before Prigogine's speech in 1995, I had never put two and two together and seen that a universal principle explained why Mother Nature and I were arriving at similar answers. The "click" I experienced that evening made me lift my eyes from my work and consider the shape and structure of everything around me. It made me wonder: What generates all these configurations? Why does this geometry happen?

I am not the first person to ask these questions. The only thing rarer in science than the eureka moment of discovery is the lone researcher who makes a discovery completely on his own. Darwin, for example, was one of many scientists exploring the evolution of species. It was his genius to imagine mechanisms such as natural selection through which evolution occurs in biology. But knowledge is not static. The human mind persistently seeks better answers to ancient questions, better understandings to ease the flow of information.

Design in nature is generating a lot of excitement today over the entire range of science—from geophysics and biology to social dynamics and engineering. The interest is fueled by two trends:

1. A voluminous body of knowledge has accumulated, and it shows that features our minds perceive as design (configurations, rhythms, scaling rules) are present in all flow systems in nature.
2. Design phenomena are not covered by the existing laws of physics.

The empirical knowledge has far outpaced the theoretical framework that is needed to support it. This kind of mismatch is the ammunition and trigger for scientific revolution. If science is an evolving animal design, then the animal has become too heavy and has no alternative but to develop a larger skeleton for itself.

From the clash between the empirical and the theoretical comes the better science, the larger skeleton that includes a law to support all the phenomena of design and evolution in nature. Many other scientists have offered their own insights into the riddle of design in nature. To varying degrees these include fractal geometry, complexity theory, network theories, chaos theory, power laws (allometric scaling rules), and other "general models" and optimality statements (minimum, maximum, optimum), as well as Charles Darwin's seminal work and D'Arcy Thompson's magisterial volume *On Growth and Form*.

My work is not a response to, or critique of, their efforts. In fact, I became acquainted with this vast literature only after discovering the constructal law in 1995. What I did know at the time was thermodynamics, the science of how to convert heat into work and work into heat. Work represents movement and flow against forces that resist. Thermodynamics rests on two laws. Both are first principles: The first law commands the conservation of energy, and the second law summarizes the tendency of all currents to flow from high (temperature, pressure) to low. These two laws are about systems in the most general sense, viewed as black boxes, without shape and structure.

Not appreciated then was that the two laws of thermodynamics do not account for nature *completely*. Nature is not made of black boxes. Nature's boxes are filled with configurations—even the fact that they have names (rivers, blood vessels) is due to their appearance, pattern, or design. Where the second law commands that things should flow from high to low, the constructal law commands that they should flow *in configurations* that flow more and more easily over time.

It occurred to me that if physics is to cover nature completely, it must be endowed with an additional first principle that accounts for the phenomenon of design generation and evolution everywhere and in everything. The constructal law is this new addition.

Previous attempts to explain design in nature are based on empiricism: observing first, thinking and explaining after. All these attempts articulate conclusions about observations that have accumulated into a body of knowledge. They are backward-looking, descriptive, and explanatory, not predictive. Darwin, for example, gathered all his observations about the evolution of biological creatures and created a convincing narrative that fit those known facts, which has been borne out by subsequent findings. Likewise, fractal geometry is descriptive, not predictive. Proponents of fractal geometry create mathematical algorithms to manufacture images that look like natural phenomena, such as snowflakes, lightning bolts, and trees. The algorithms they devise in order to draw these images are not derived from principle but from trial and error. The algorithm that the mathematician chooses in order to draw the tree in the garden is analogous to the brush and paint that the painter chooses in order to depict the same object. The mathematician shows us only the algorithms and drawings that come out right, not those that look like nothing. The painter does the same.

The constructal law does much more than explain the designs we see in nature. It articulates a law we can use to understand *why* designs emerge and predict *how* they will evolve in the future.

In recent years, many members of the scientific community have begun questioning the strictures of Darwin's work—combating what the biologist J. Scott Turner has called, in his 2007 book *The Tinkerer's Accomplice*, the "pernicious tendency for the convenient assumption to become unquestioned dogma."

Not surprisingly, evidence of design in nature has sparked this robust inquiry. While delivering the prestigious 2007 Gifford Lecture at the University of Edinburgh, the British paleontologist Simon Conway Morris argued that evolution shows an eerie predictability, leading to the direct contradiction of the currently accepted wisdom that insists on evolution being governed by the contingencies of circumstances.

And Turner has observed "a peculiar harmony of structure and function in the devices organisms contrive to accomplish things." Natural selection, he continues, cannot fully explain this because it is "contingent upon the past but with no view to the future, and with certainly no purposefulness or intelligence guiding the process."

In an interview at Brown University in advance of his appearance at a 2008 symposium of the American Association for the Advancement of Science in Boston, Brown biologist Kenneth Miller said, "The idea that there is 'design' in nature is very appealing. People want to believe that life isn't purposeless and random. That's why the intelligent design movement wins the emotional battle for adherents despite its utter lack of scientific support. To fight back," he continued, "scientists need to reclaim the language of 'design' and the sense of purpose and value inherent in a scientific understanding of nature. . . . There is, indeed, a design to life—an evolutionary design. The structures in our bodies have changed over time, as have its functions. Scientists should embrace this concept of 'design' and, in so doing, claim for science the sense of orderly rationality in nature to which the anti-evolution movement has long appealed."

Morris, Miller, Turner, and others have the right hunch: Design in nature does not arise by accident. Their comments underscore the fact that we are living in revolutionary times, when fundamental assumptions are being challenged. But most scientists are willing to go only so far with their iconoclasm. Even as they question some tenets of Darwin and his followers, they hold on to the idea that biological organisms are different from everything else. The celebrated science writer Richard Dawkins articulated this view in his acclaimed book *The Blind Watchmaker: Why the Evidence of Evolution Reveals a Universe Without Design*, when he asked "how [complicated things] came into existence and why they are so complicated." He argued, "the explanation . . . is likely to be broadly the same for complicated things everywhere in the universe; the same for us, for chimpanzees, worms, oak trees and monsters from outer space." Just when it seems he is going to offer a universal outlook, he pulls back. "On the other hand, it will not be the same for . . . 'simple' things, such as rocks, clouds, rivers, galaxies and quarks. These are the stuff of physics. Chimps and dogs and bats and cockroaches and people . . . are the stuff of biology."

This fundamental division between physics and biology is false. It does not result from a broad view of how the world works but from that ancient adage: Your answers are only as good as the assumptions underlying your questions. Darwin and his followers heroically helped remove God from the scientific equation. And, to the discomfort of many, they took human beings down a peg or two when it comes to our place in the cosmos. But they couldn't completely break from the past, couldn't see beyond the idea that biological life is special.

The remnants of this old worldview aren't the only things that have hindered understanding. At its best, science encompasses everything—it seeks to provide a rational basis for all that is. However, especially during the last two hundred years, its practitioners have tended to slice and dice the universe into smaller and smaller pieces, all the way to the infinitesimal. Some people study rocks, others look at birds; some study space, others focus on human beings. You may have noticed the same phenomenon when you seek medical treatment—one doctor specializes in kidneys, another in colons, another in the heart; no one can manage all your care.

Because scientists have focused on ever-smaller questions, and ever-smaller dimensions, most have failed to see the big picture. This has prevented even those who are aware of the overarching tendencies of design in nature from taking the imaginative leap to see that the broad evolutionary tendencies we observe in living creatures also shape inanimate phenomena that do not possess DNA subject to random mutation, such as rivers, global weather patterns, and everything else that moves.

I took this step in 1996. While writing my second paper on the constructal law for an international journal, I noted:

> A lot has been written about natural selection and the impact that thermodynamic efficiency has on survival. In fact, to refer to living systems as complex power plants has become routine. The tendency of living systems to become optimized in every building block and to develop optimal associations of such building blocks has not been explained; it has been abandoned to the notion that it is imprinted in the genetic code of the organism.
>
> If this is so, then what genetic code might be responsible for the development of equivalent structures in such non-living systems as rivers and lightning? . . . Whose genetic code is responsible for the societal "trees" that connect us, for all the electronic circuits, telephone lines, air lines

> [routes], assembly lines, alleys, streets, highways, and elevator shafts in multistory buildings?

I am not disputing the role of genetics in the origin of species—just as I don't discount the pivotal role of soil erosion in the formation of river basins. But mechanism is not law. It may explain what has happened but not why it should happen. Indeed, in view of the constructal law, we see that the search for mechanism has been monumentally unproductive for the understanding of design in nature. There is no single mechanism that generates design in river basins and biological organisms. Instead there is a single principle of physics that governs the design-generating action of soil erosion or genetics. On one level these two phenomena couldn't be more different—yet both create shape and structure that facilitate flow. Natural selection, random mutation, and soil erosion are not the endgame. They are just three of the many morphing mechanisms we find in nature that serve the unifying principle for all evolutionary phenomena, the constructal law.

The constructal law also challenges another idea that has become dogma since Darwin—that there is no overarching direction to evolution. Proponents of that view claim that adaptations make species better able to survive, but they never explain why these changes should occur and what they mean by "better." The closest they come is through a piece of circular logic that says: A change is better if it aids survival; any change that aids survival is better. The constructal law, by contrast, predicts that evolution should occur because of the tendency of all flow systems to generate better and better designs for the currents that flow through them. It expresses the meaning of "better" in unambiguous physics terms—change that facilitates faster, easier movement. As we will see, not only do river basins and forests improve in time but so do biological creatures—the rise of species from single-cell organisms to fish, birds, and humans is the story of better, more efficient flow of animal mass on the landscape. In big history, all these designs have emerged because they enhance the movement, mixing, and churning of energy and mass on the planet.

The constructal law is the latest advance in our ever-evolving understanding of nature. Yet, on a basic level, my work is connected to those who came before, both in and out of science, who have tried to describe the flowing world around them. The novelist William Faulkner, for example, hinted at my new definition of life when he wrote, "'living' is motion, and 'motion' is change and alteration and therefore the alternative to motion is un-motion, stasis, death. . . ."

Although Faulkner spoke of human beings, not rivers, his quote suggests that people have long understood a basic truth of the constructal law that is encompassed by the old sayings "going with the flow," "taking the path of least resistance," and "doing the most with the least." The American transcendentalist Henry David Thoreau expressed this as a philosophy of life in 1853 when he wrote: "Dwell as near as possible to the channel in which your life flows." The nineteenth-century American economist Henry George articulated this principle as well when he observed: "The fundamental principle of human action . . . is that men seek to gratify their desires with the least exertion."

The idea that nature organizes itself to move more easily has a long pedigree in the sciences, too. In the first century CE, Heron of Alexandria intuited that a ray of light bouncing off a mirror and traveling between two points follows the shortest path. From this mental viewing he predicted the shape of the reflected ray, that is, that the angle of incidence should be equal to the angle of reflection. In the seventeenth century, Pierre Fermat had a similar vision, the concept of minimum travel time, when he predicted the shape of the refracted ray, that is, the broken ray when light passes from air into water.

The great scientists who developed mechanics and calculus three centuries ago (Newton, Leibniz, Euler, the Bernoullis, Maupertuis, Lagrange) began to question design in nature by thinking that nature optimizes things. Variational calculus emerged as a technique for identifying "optimal" paths—ultimate drawings, "destined" to satisfy specific objectives when constraints (aka reality) are also taken into account. Close, but no cigar. Nature does not produce optima, or "end designs" or "destiny." Nature is governed by the tendency to generate shapes and design that evolve in time to reduce imperfection. Design evolution never ends.

The constructal law is not about destiny (or optimum, maximum, minimum, most, least, best, worst, etc.). Yet the insights from the eighteenth century suggest one of the powers of the constructal law: It offers a scientific confirmation, a rational, testable basis for our intuition that there is a direction in time to the evolution of all around us, a purpose, a direction toward flow performance in all that goes on around us.

The constructal law also helps us see another fact that people have long intuited—the harmony in nature. Rivers are lovely for many reasons, and one of them is that they follow geometric rules predicted by the constructal law: Their depth is proportional with their width—big streams are wide and deep; small streams are narrow and shallow. This, of course, is good for the flow of water. This and the myriad other scaling laws we find in nature are only surface reflections of a far deeper harmony. As we will see, our ideas of beauty take practical form when we see how they are often reflected in natural designs.

The constructal law teaches us that nothing operates in isolation; every flow system is part of a bigger flow system, shaped by and in service to the world around it. The flow system we call a tree is also part of the larger flow system (that also includes rivers and weather patterns) for moving water from the ground to the air in order to achieve an equilibrium of moisture locally and globally. At the end of the day, the tree, like every other flow system, exists in order to facilitate nature's tendency to flow with configuration. Its shape and structure reflect the tendency to generate designs to do this efficiently. This interdependence, born of thermodynamics and the constructal law, is the true source of harmony, balance, and oneness in nature.

In my professional capacity I see the constructal law as a powerful scientific tool. As a human being, I also appreciate its metaphysical implications. Poets have long celebrated the balance and harmony of the world, the oneness of nature. But this has been hard to prove rationally. Until now. By identifying a principle that joins the animate and inanimate worlds, that links the flow of rivers to the flow of cities and the flow of money, the design of our lungs and blood vessels to trees and lightning bolts, the constructal law brings science in line with poetry. It reveals our deep connection. It illuminates the tendency that unites everything that moves.

CHAPTER 1

The Birth of Flow

Not many outsiders study the Romanian language, so those who are born into it have few opportunities to use it beyond the country's borders. On the plus side, Romanian is a very special Romance language—a highly preserved form of Latin—and it is very similar to Italian. This, and the fact that the many invaders who have flowed in and out of Romania during the centuries exposed the language to Greek; Slavic; Germanic; and Asiatic, non-Indo-European elements (Hungarian, Tatar, Turkish, in this order), have made it easy for Romanians to understand many modern languages. My French, for example, was good enough that I was able to understand Ilya Prigogine's speech and his assertion that day in France that the treelike structures that abound in nature are the result of chance.

Had I been reared speaking English, perhaps I would have not understood Prigogine. I am more certain that had I not become an American—and the access to people, places, and ideas that affords—I never would have discovered the constructal law, because I would not have had the opportunity to be present in the room where Prigogine spoke.

I spoke almost no English when I arrived in America in February 1969 to study at MIT. Like most immigrants, I learned

English on the fly. Because of my origin, I have always paid particular attention to the history and precise meaning of words. Precision is the foundation of both language and science. The definition of a word, like the boundaries of a thermodynamic system, spells out exactly what it is and what it is not. Science and language are rooted in the past, which means both history and geography. Neither is static or appears out of thin air. Both come from somewhere and evolve in time, building on all that has come before, providing, like the channels of a river basin, better access for the currents that flow through them.

My amateur interest in language took on a more decided purpose sixteen years ago after my discovery in Nancy. Before then, my work as a professor of mechanical engineering involved applying the laws of thermodynamics, mechanics, and heat and fluid flow to practical problems of heat transfer and cooling. If you wanted to build a better computer, refrigerator, or power plant, I was your guy.

When I noticed that the systems I was creating were strikingly similar to those that appear in nature—and that a single principle of physics, the constructal law, accounts for their design—I was thrust into a strange and contentious world. In this scientific Tower of Babel, I learned that the common words I used to describe my discovery—especially "evolution," "direction," "purpose," and "design"—were weighted with history and fraught with controversy. Instead of my computer, I needed a dictionary.

Start with the word "design." Its meaning seems straightforward. It is the configuration, transformation, or assembly of materials with a specific purpose—taking something today and intentionally changing it so that it will be something else tomorrow. On the one hand, this is one of the most obvious and unchallenged concepts known to humanity. The modern world is built by the simple process of turning raw materials, such as metals and minerals, plants and animals, into useful things. Look around your house. Someone designed everything in it, from the building itself to the pipes that bring water to your sink, to the appliances that sit on your kitchen counter to the money you carry. So, too, were the clothes on your back and the art on your walls. Billions of people earn their livings creating and constructing designs.

Design may be the foundation of the built world, but it is anathema when the conversation turns to nature. Its six letters have become the four-letter word of biology and physics. If you want to send a chill across a lecture hall full of scientists, just mention design in nature. If we claim that rivers, trees, or snowflakes reflect design, the question naturally arises: Designed by whom, for what? For thousands of years, people of varying faiths answered this question with ease. Divine forces created the shapes and patterns in nature. The gods, either one or many depending on the era or belief system, were the Master Builders.

As the Renaissance and the Enlightenment flourished, rational minds searched for evidence of this claim. Creationists and defenders of what we now call intelligent design didn't produce any ironclad proof. They asserted that natural designs were so intricate and complex, that they exhibited so much order and direction, that they could not have resulted from blind forces. The most famous articulation of this teleological argument—from the Greek *telos*, meaning "end" or "purpose"—was offered by the British thinker William Paley in his 1802 book, *Natural Theology: or, Evidences of the Existence and Attributes of the Deity*, where he likened God to a watchmaker:

> In crossing a heath, suppose I pitched my foot against a *stone*, and were asked how the stone came to be there; I might possibly answer, that, for any thing I knew to the contrary, it had lain there for ever: nor would it perhaps be very easy to show the absurdity of this answer. But suppose I had found a *watch* upon the ground, and it should be inquired how the watch happened to be in that place; I should hardly think of the answer which I had before given, that, for any thing I knew, the watch might have

> always been there. Yet why should not this answer serve for the watch as well as for the stone? Why is it not as admissable in the second case, as in the first? For this reason, and for no other, viz. that when we come to inspect the watch, we perceive (what we could not discover in the stone) that its several parts are framed and put together for a purpose. . . . This mechanism being observed . . . the inference, we think, is inevitable, that the watch must have had a maker: that there must have existed, at some time, and at some place or other, an artificer or artificers who formed it for the purpose which we find it actually to answer; who comprehended its construction, and designed its use.

About a half century later, Charles Darwin seemed to deliver the deathblow to this line of thinking. Confining himself to biology, he argued that the appearance of design we see in complex life-forms does not reflect divine intent. It results instead from the mindless process of evolution by natural selection, the "principle," he wrote, "by which each slight variation [of a trait], if useful, is preserved." A bird, for example, is not assembled at once—each part positioned just right in relation to all the others—like a watch. Instead, it has emerged through an evolutionary process with no larger direction or purpose. Small adaptations that provide some advantage make certain species more fit to survive, to reproduce, than others. A finch that lives in an environment where tasty seeds are somewhat hard to reach will do better if it has a longer beak. Those that have longer beaks survive and pass the trait on to their descendants, who continue to evolve.

The finch cannot will itself to have a longer beak. The mechanism by which this occurred remained a mystery until the Austrian monk Gregor Mendel showed through his famous experiments with pea plants how traits are inherited. His work led to the modern science of genetics, so that it is now believed that random genetic mutations create different traits. This process occurs ceaselessly. Sometimes, these changes produce beneficial results, oftentimes not. In general, the theory goes, when helpful variations appear, they tend to stick.

Darwin, Mendel, and others have provided us with tremendous insights that have led to innumerable benefits. More than just a scientific hunch, they have given us a deeply entrenched worldview born out of battles with old ideas. My guess is that most readers of this book have been schooled in its assumptions, embrace its meanings, and speak its language. At best, I was vaguely familiar with their work as I pursued my own. But it was also the furthest thing from my mind as I worked on the constructal law. So when I started speaking of design in nature, of the direction and purpose of evolutionary phenomena, I found myself embroiled in long-simmering controversies. I felt that people wanted me to use old language to describe a new way of understanding. Besides, I had no choice in the matter: Discovery precedes language; it takes time for language to evolve to facilitate the flow of new ideas. This meant that I had to use existing terms to describe my work.

I share this brief history with you because it has defined the terms of the debate I have entered into but want no part of. The constructal law is not a response to these claims but a different way of defining and understanding the concepts it evokes.

It claims no more and no less than this: Everything that moves is a flow system that *evolves* over time; design generation and evolution are universal phenomena. The changes we witness in animals, plants, rivers, and steaming pots of rice represent a clear improvement over the configuration that had been flowing before. This is the *direction* of evolution, creating flows that move more easily, better, farther, etc. The *design* we see in nature—the shape and structure of rivers, animals, cities, etc.—is a manifestation of this tendency in nature to generate shape and structure to facilitate flow access.

This is direction and evolution without intention. Flow systems do not *want* to move more easily; they do not *seek* greater access for the currents that flow through them. They do so because they are governed by the principle of physics described by the constructal law. I know this is a hard idea to grasp—design without a designer. It becomes easier when you consider another law, gravity. If you go to the top of a building and drop a rock, it will fall, faster and faster. No one would argue that the rock wants to fall, but it does, it must.

At a basic level, science is the search for such laws—simple, efficient statements of the impersonal and predictable tendencies of nature. Its knowledge turns us into sorcerers and soothsayers, able to tell the future with certainty: If X occurs, Y will happen: Water will boil at a certain temperature; a steel bar will bend if enough pressure is applied to it; I can swim in water but not in the air. Intentionality has nothing to do with any of this. These are the rules of the road that must be obeyed. Imagine if there were no predictability or order in nature. If water had a random rather than a predictable boiling point, it might take a second for you to make tea or an hour or eternity. If steel didn't have a predictable breaking point, we could never build safe houses or cars. Life as we know it would be impossible if we couldn't count on the laws of physics.

The constructal law identifies a phenomenon as old as the universe itself but unrecognized until now. Its power and correctness rest on this fact: It enables us not just to describe but to predict the evolution of all flow systems. Before making any observations in nature, we can use the constructal law to imagine what a lung, blood vessel, tree, river, or lightning bolt *should* look like if it has the freedom to change over time to flow more easily. When we compare our drawings with what we find in the real world, they line up.

The constructal law predicts that for rivers to flow more efficiently, their width should be proportional to their depth; that the circulatory system in our bodies should have a treelike structure of round tubes with a few main channels (arteries and veins) and numerous tributaries (capillaries) in order to deliver water, oxygen, and useful energy to every cell; and that our hearts should have an intermittent beat—lubb-dupp, pause, lubb-dupp, pause—because that is an efficient way to deliver oxygen and other substances to the living body.

To see how, let's take a closer look at our respiratory system. Its primary function is to draw oxygen from the air by inhaling, oxygenate our blood in the tiny alveoli of the lungs, and then drain carbon dioxide from the blood and then out of our bodies by exhaling. Instead of studying the system we find in nature, we use the constructal law to imagine what a theoretical fluid flow architecture should look like if it is to perform these functions with low mass transfer resistance and low fluid flow resistance, all in a small volume. (This is quite a mouthful! I thought about this when I wrote the constructal law, and I condensed it all into a statement of design and evolution toward greater "access" for all the components in a "finite-size" system.) In this case, we ask: What design would we come up with to promote access to the external air while also bathing the entire volume of the lung with oxygen?

Using pencil, paper, and the constructal law, Professor Heitor Reis and his associates at the University of Évora in Portugal demonstrated that the best oxygen access to the alveolar tissue is provided by a treelike flow structure composed of ducts with twenty-three levels of bifurcation (that is, they continuously double the number of tubes as they split) that ends with alveolar sacs from which oxygen diffuses into the surrounding tissues. Because mice, for instance, are smaller, they should (and do) have nine levels of bifurcation in their lungs.

Among other things (including the dimensions of the alveoli and the total alveolar surface), Reis and his associates determined that good flow access is achieved when the length of each bifurcating airway (the two daughter airways that branch off from each mother airway) is defined by the ratio of the square of the first airway diameter to its length.

When they compared their findings to our actual respiratory system, they found that they had predicted its shape and structure: twenty-three levels of bifurcation with a fixed ratio in the length of mother and daughter airways.

In my own work, I tackled a different aspect of the respiratory system, the rhythm of breathing. I asked: What would be a good rate for inhaling and exhaling to deliver oxygen and remove carbon dioxide? I found that resistance is decreased when there are intervals and that these intervals should be proportional to the mass (M) raised to the power of 0.24. This means that smaller animals should breathe more frequently than larger ones, which is in agreement with voluminous empirical observations.

Physiologists tend to focus on the breathing rhythm, and many have proposed empirical models for summarizing quantitatively the observation that large animals breathe less frequently. By starting with the constructal law, I discovered three other features that are so obvious that nobody had questioned them before:

1. The flow architecture of the lung could have any design, yet it is shaped as a tree. This is important because today the lungs and other organs are modeled routinely by postulating a fractal-like design because the air passages in the lungs look like a tree. The question should have been: Why should it be treelike? Why not just a single wide air pipe?
2. Breathing could have any design. In fact, the design that would require the least effort would be a single breath that lasts a lifetime. But breathing is a periodic flow, inhaling and exhaling. The question should have been: How should the airflow vary in time? The notion of periodic flow (inhaling, exhaling) does not exist theoretically before this question is posed and answered.
3. The inhaling timescale must be the same as the exhaling timescale, regardless of the animal's size. Mice have equally short inhaling and exhaling times. Cows have equally long inhaling and exhaling time intervals. So

> do people, whether we are breathing at a relatively slow rate while sitting in a chair or quickly while exerting ourselves.

Of course, most phenomena do not exhibit this splendid predictability of lung structure. Design in nature is not precise to the nth degree—otherwise every tree would be identical. The diversity we find is immense. Phylogeny (the study of the evolutionary relatedness among various groups of organisms) creates genetic architectures that, like boulders in rivers, are slow to erode. Thus, many of the "mistakes" biologists point to in animal design—for example, the long, winding path taken by the recurrent laryngeal nerve in mammals when a shorter path would be more ideal—reflect the move toward better flow within the constraints of evolutionary history. In the inanimate world, ecological variation leads to other anomalies. Multiple factors shape design, just as, for example, constant exposure to a high wind and the input of surrounding sediment would cause a different form of river basin to evolve by processes we ordinarily address without reference to the wind or an exhaustive list of other factors that create idiosyncratic variation.

What the constructal law captures is a central tendency in nature. In the big picture—all the rivers, all the trees, all the animals on Earth—the driving force behind the evolution of everything that flows is the generation of shape and structure to move more easily. This is why we can say that this unintentional tendency has a *purpose*. All the flow designs on Earth, from air currents, rivers, and trees to fish, people, birds, and technology arise, evolve, and compound themselves to enhance movement. It is because this tendency has a direction, a purpose, that we can predict how things should evolve in the future.

Using the constructal law we can recast organic evolution as a dynamic process that generates better designs. There is a large volume of imperfection involved (genetic drift, selection on linked alleles, extinction, dispersal limitation, environmental heterogeneity in space and time, etc.) as well as idiosyncratic variation. But the central tendency is the selection of characteristics that ease the flow—that allow animals and plants to generate more flow (movement) for their mass with less expenditure of useful energy to achieve this movement. The evolutionary history biologists have charted, the series of adaptations they have detailed, are all expressions of the constructal law. The same tendency governs the evolution of inanimate systems; of all the possible configurations, the ones that persist are those that facilitate flow. Thus, the constructal law shows that animate and inanimate entities develop flow configurations in a manner consistent with the idea of natural selection in biology. The dynamic processes and features of inanimate phenomena, such as river systems, can be united via the constructal law with the dynamic features that inform the evolution of biological organisms including you and me.

One reason that I was able to discover the constructal law is that I was not immersed in the language and history of Darwinism. My field is thermodynamics, and it is from here that my language and insights developed. Through it I identified the principle that generates design in nature.

To have a conversation, we have to speak the same language. Otherwise, we are not communicating. So I turn to thermodynamics to define some basic terms and address some of the deeper questions raised by the constructal law: What is a flow system? Why do things flow? Why do they evolve? How are direction, purpose, and design possible without the guiding hand of God or man?

The field of thermodynamics was born because of the Industrial Revolution. It can be said to begin with the great French scientist and inventor Denis Papin (1647–1712), who was fascinated by fire and water. One of his earliest inventions was what he called a "steam digester, or engine for softening bones." Observing that the pressure inside his cooker could be harnessed for other uses, he devised the first piston and cylinder mechanism, a design that still powers our engines. In his primitive yet practical device, high-pressure steam is created by heating water in a confined space. The steam pushes a piston that resists the pushing, and that means the steam does work on the piston. In modern turbines, such chambers are formed repeatedly between

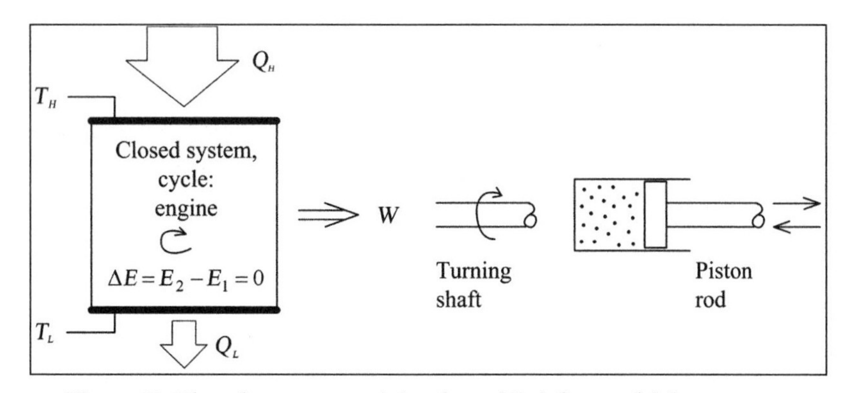

Figure 7. Closed system receiving heat (Q_H) from a high-temperature reservoir (T_H), rejecting heat (Q_L) into a low-temperature reservoir (T_L), and producing work (W). The energy inventory of the system (E) changes from E_1 to E_2. The first law of thermodynamics states that energy is conserved. The energy that flows in (Q_H) minus the energy that flows out ($W + Q_L$) equals the amount of energy stored inside the system [$E_2 - E_1 = Q_H - (W + Q_L)$]. If the system executes cycles, then at the end of each cycle, all the system properties return to their original values; for example, $E_2 = E_1$. A real engine can always be improved, because its work output (W) is always smaller than its largest theoretical value (W_{rev}), which occurs in the ideal limit of reversible operation. In this ideal limit imagined by Carnot, all the currents and motions of the engine system occur in the absence of resistances such as friction and heat leaks. Any real engine is completely equivalent to an ideal engine producing maximum work (W_{rev}) and destroying a portion of this work (W_{diss}) into a brake. It is as if the shaft of the ideal engine is resisted by the brake but only partially, because a portion of W_{rev} (namely W) is eventually delivered to the environment. In the limit where no user exists in the environment to receive the work (W), all the work produced by the engine is dissipated into the brake. This limit represents the design of everything that flows and moves on Earth. (We return to this natural design in Figures 57 and 59.)

rows of blades rotating against rows of stationary vanes, with steam (or another hot gas) temporarily trapped in these chambers as it expands (Figure 7).

In summary, we create work from heat, as in the name "thermodynamics" coined for this science by the nineteenth-century Scottish physicist William Thomson (later Lord Kelvin)—*therme* ("heat" in Greek) and *dynamis* ("force"). No animals or slaves are needed to push the body out of the way. Is any idea cleverer than this? Is any idea more humane?

The laws of thermodynamics apply to everything on Earth. All natural designs are engines (of heat, fluid, or mass) driven by useful energy derived from the sun. All things that flow, including people and other biological creatures, acquire evolving designs that allow them to move more current farther per unit of useful energy consumed.

Today thermodynamics encompasses all aspects of energy and energy transformation, including refrigeration (which involves the removal of heat from a system colder than the ambient) as well as the relationships among properties of matter and power production, from the operation of power plants to photosynthesis.

As the study of thermodynamics grew, its practitioners developed a vocabulary that allowed them to know precisely what they were talking about. One of the most basic terms is the word "system," which means the region in space or the quantity of mass selected by an observer for analysis and discussion. A tree is a system we might study, but so is the forest of which it might be part. How we define the "system" is a personal choice, but it must be placed naked on the table. We define the system by drawing a sharp and precise boundary around the entity in question (Figure 8).

Inside the boundary is the system we have decided to look at—the tree, the forest, the stream, the river basin. Outside the boundary is everything else: the rest of the world. This "other system" in thermodynamics is called the "environment," or the "ambient." The water inside Papin's primitive cylinder was a sys-

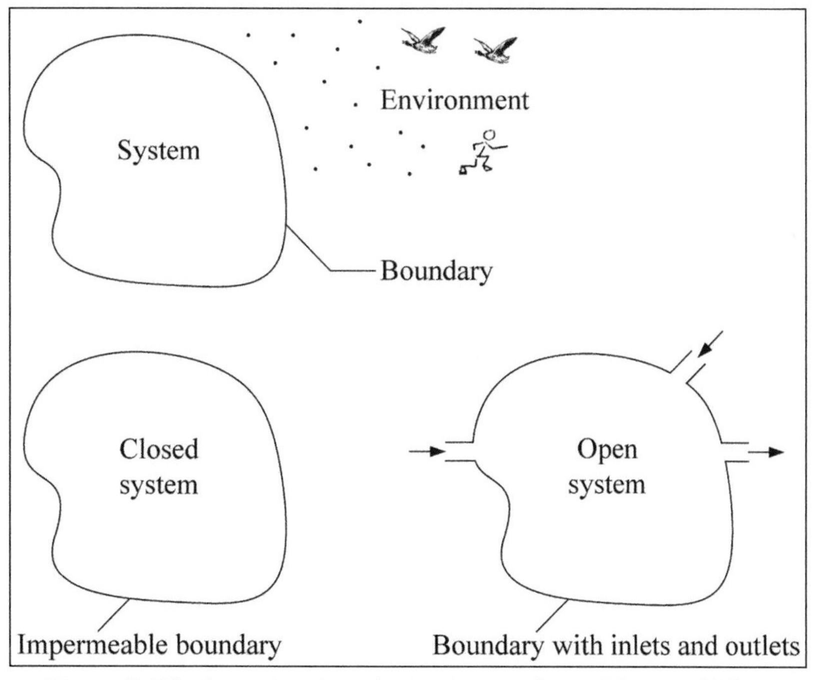

Figure 8. The boundary is an imaginary surface with zero thickness that separates the system from the rest of the universe.

tem for creating steam. A river basin is a system through which water flows; a bird is a system that moves the animal's mass across the Earth.

The boundary can have special features, which give the system special names. If the boundary is impermeable to the flow of mass, then the system is called "closed" and its mass inventory is fixed even in cases when its state, its "being," is unsteady. Think of the system represented by the air inside an impermeable balloon. Because there is no leakage of air from the balloon to the environment, the amount of air stored inside the system is fixed.

If mass crosses the boundary, the system is called "open." Speaking thermodynamically, your body is an open system. Its boundary is an imaginary surface drawn on the outside of your skin—a three-dimensional version of the chalk outline detectives draw at murder scenes. This boundary has inlet ports (the mouth and nose) that introduce substances into the system and outlet ports (including the mouth and nose) as well as skin pores and other organs that leak substances into the environment.

Mass goes in, mass goes out, and the system itself can move in space (e.g., the bird, or vehicle). The key point is that mass is conserved: The difference between input and output is the mass that is being stored (accumulated) inside the system. We have just described a law obeyed by all systems, open or closed: the law of mass conservation: The whole mass must be equal to the sum of its parts. Though this law was discovered by Antoine Lavoisier in the eighteenth century, the ancient Greek philosopher Epicurus hit on this idea when he postulated that "the sum total of things was always such as it is now, and such it will ever remain."

A similar law applies to energy. It, too, can neither be created nor destroyed. The law of energy conservation is known as the first law of thermodynamics and was articulated (at the same time as the second law, which we will discuss next) by three men from 1851 to '52—two Scotsmen, William Macquorn Rankine and William Thomson (later Lord Kelvin), and the German Rudolf Clausius. This, much more than the romantic ideal of the lone genius, is usually how science works; an inkling arises that there is a better way to speak of nature, inspiring researchers to try (often in competition with one another) to discover and articulate it. We see this time and again in the history of science in general and the constructal law in particular: Early practitioners observe a phenomenon, but they do not possess the knowledge to predict it. In Darwin's time, others were working to discover the mechanism of biological evolution; today, many others are examining the phenomenon explored in this book: design in nature.

Building on the visionary ideas of the Frenchman Sadi Carnot (1796–1832), Rankine, Kelvin, and Clausius observed that while a system cannot create energy, it will conserve and can transform it. Our cars transform the useful energy of gasoline into heat that drives our engines; the engines partially convert that heat into the work needed to push our cars on the road. Our bodies convert the energy from food to power ourselves. Power plants transform energy from one type (heat) into another type (work). When we examine any system, we ask: How well is it doing? How efficiently is it using the available useful energy? In thermodynamics the usual way of measuring this progress is by speaking of the "energy-conversion efficiency" of the engine. Better designs are more efficient; they perform more work for less useful energy.

This concept is key when we speak of flow and flow systems. Flow represents the movement of one entity (in the channel) relative to another (the background). To describe a flow we speak of what the flow carries (fluid, heat, mass, information), how much it carries (mass flow rate, heat current, traffic, etc.), and where the stream is located. Flow systems are defined as one or more streams that originate from points and must find easier access to other points.

The constructal law predicts that flow systems should improve over time. This raises the question: Better in terms of what? How do we measure improvement in the system? The answer lies in the fact that flow systems are imperfect thermodynamically because of the resistances their flows must overcome. Resistance is the phenomenon of opposition to movement. It is best known as friction—for example, the horizontal force that a human, animal, or motor must exert on a vehicle in order to make it move horizontally. Fluid friction acts in the same direction: a pump must maintain a high enough pressure at the entrance to a long pipe in order to force the column of water to move along the pipe. Much more subtle is the "friction" encountered and overcome by a heat current, which must be "driven" by a finite-size temperature difference in order to flow. This heat-flow kind of friction is thermodynamically equivalent to the mechanical friction.

Imperfections are unavoidable. In fact, they are necessary. Without imperfections (resistances), flow systems would accelerate continuously, eventually spinning out of control. Thus, imperfection (friction, heat leaks, etc.) acts as a brake on the engines (the designs) that drive flow. I know this firsthand from my own work. Like all engineers, when I set out to design a device or a system, I have to understand the function it will perform and the hurdles that stand in the way. I shape and assemble its parts so that the global system will function in the least imperfect manner possible. My struggle as a designer never ends.

So it is with all natural designs, which encounter various forms of resistance—choke points, bottlenecks, friction, drag, thermal insulation, etc. To cite one common example, imagine that you are on the banks of the River Thames. Don't forget your jacket, because it's a typical spring day in England—gray, dank, a bit nippy. Still, the air is filled with excitement because it's Race Day; the oarsmen of Cambridge are poised for battle against their adversaries from Oxford. As the powerful combatants settle into their boats, they appear to be preparing to compete against each other, but their true foe is the water itself. The legendary Cambridge rower Steve Fairbairn immortalized their mighty motions in "The Oarsman's Song":

The willowy sway of the hands away
And the water boiling aft,
The elastic spring, the steely fling
That drives the flying craft.

That has a little more zip than my analysis of the oarsmen's efforts, but I will try to make up in precision what I lack in grace. Driving the boat requires an expenditure of work (W) from the rower pulling the oars. The work is spent in order to overcome the friction force (F) due to two effects: friction as the boat slides on the water to cover a distance (L) and the need to lift the water, to get it out of the way. This second effect is visible as waves. The work spent is the resistance force times the displacement ($W = FL$). More friction force requires more work to cover the same distance. This is why the leading boat will move as close as possible to its competitor's lane. The momentum of the water that the rowers push behind them creates even more resistance for their adversaries to overcome, so that they must do even more work to cover the same distance. It also explains why today's oarsmen do not use the same boats as their ancestors in that first contest in 1829. Through the years, craftspeople have built ever more hydrodynamic boats to minimize the effects of friction. Better designs lead to greater efficiency, which means easier movement. This is part of the template for all design evolution in nature.

Fish, land animals, birds (that is, all swimmers, runners, and fliers) are flow systems that move mass (themselves) on the globe. The food they take in provides the useful energy that allows them to move across the landscape. To move they must overcome two obstacles—the downward pull of gravity and friction from the water, land, and air. As we will see in chapter 3, their design (which includes everything from the shape and structure of their innards to their total body mass and the rhythmic motion of their tails, legs, or wings) has evolved to allow them to move more easily and efficiently in an environment that resists the movement.

Finally, consider the maple tree. It is not just a source of shade for romantic picnics but is also a flow system for moving water from the ground to the air. While battling gravity and friction to move water up its length, the tree must also stabilize itself against the resistance caused by the wind, which would sever its limbs and topple it over. Through the constructal law we find that every aspect of its design—the shape of its roots, trunk, branches, and leaves—can be predicted when we recognize the two flow systems in the maple: water and stresses. The standing tree is facilitating the movement of water and mass on the globe, as we will see in chapter 5.

This is what we see in every flow system. The road to easier flowing consists of balancing each imperfection against the rest. All the components of the system collaborate, working together to create a whole that is less and less imperfect in time. The distributing and redistributing of imperfection through the complex flow system are accomplished by making changes in the flow architecture. This is true for all animate and inanimate designs, from the placement of branches on a tree to the channels in a river basin to the arrangement of electronics in your laptop.

A prerequisite, then, is for the flow system to be free to morph. The emerging flow architecture is the means by which the flow system achieves its objective under constraints. *Freedom is good for design.*

Before leaving this discussion of thermodynamics, we have to explore one more crucial point as it relates to flow. Why does anything flow at all? Why does anything move? What is the wellspring of this action? For things people make, there is a simple answer—we provide fuel (i.e., useful energy, exergy) to power the devices and systems we build. But what about everything else? One obvious answer is gravity. It pulls the water from the mountaintop to the plain and sends it barreling down the river. But what puts the water on the mountaintop in the first place?

To find the answer, we return to the work of Carnot. A military engineer felled at a young age by cholera, he was a graduate of the famous École Polytechnique and later spent time at the august institution that features a statue of Papin, the Conservatoire National des Arts et Métiers (CNAM) in Paris. Carnot came to the CNAM in the early nineteenth century to study the steam engines that were transforming Europe. Why were steam engines so popular? Because their effect on people's lives was very good. It was dramatic. Engines were empowering people. They were facilitating the flow of people and goods across the globe.

As he contemplated this parade of machines, Carnot saw that everything flows in one direction: from high to low. Water flows spontaneously from high pressure to low pressure. Water falls through (and turns) a water wheel by flowing from high to low. He then reasoned that heat flows of its own accord from high temperature to low temperature. Similarly, heat falls through (and turns) the engine by flowing from high temperature to low temperature (the ambient). This "one-way flow" principle is known today as the second law of thermodynamics, irreversibility, dissipation, inefficiency, water under the bridge, etc. By looking at the operation of machines, Carnot discovered a law of nature.

The second law has been stated in several other ways. For example, if the system is *isolated* (not touched by anything)—and this is a big if—nature tends to efface differences and create uniformity. The simplest example is a glass of ice water left overnight on the kitchen table (the "isolated system" here is the entire room with the glass of water in it). In the morning the water will be at room temperature. In addition, there will be less water in the glass because the drier air has been removing water from the glass in order to reach an equilibrium of vapor pressure.

To see this action on a global scale, let's take a trip to the warmer latitudes near the Equator. The heating there is more intense because the sun's rays are almost perpendicular to the Earth's surface. Think of the rays as a bundle of arrows. Near the poles, they land almost at a slant, and this means very few of them land per unit of ground area. The uneven heating of the Earth sets in motion the ocean currents and global weather patterns. The warmer water and air head toward the poles, to mix and churn with the cold water and air there in order to create uniformity. But the route to this never-reachable uniformity is paved with nonuniformity: sharp currents, rivers, and winds that are distinct from their unmoving surroundings. This nonuniformity facilitates the global flow, and, if the sun were to disappear, accelerates Earth's system toward uniformity (death).

The sun, however, does not disappear, so the globe is not an isolated system. In the simplest description, our planet is a closed system. The sun heats the Earth and continuously forces the oceanic and atmospheric currents to keep on flowing. It forces the nonuniformity to persist. It uses the currents themselves (what moves is distinct from what does not move), illustrating for all of us that nonuniformity reigns as patterns, configurations, and rhythms. If the system is forced to flow, far from equilibrium, it generates plenty of differences, namely the heterogeneity distributed throughout the volume (or area) as channels and interstices: faster, easier flow through the channels (such as water in a river) and slower, more difficult flow in the interstices (such as the water that seeps from the ground to the river).

Let's stay in the ocean to find more evidence of spontaneously occurring design and to solve one of the great riddles of fluid mechanics. Scientists have long known that flow regimes change—from a smooth laminar flow to the whirling eddies of turbulent flow, and vice versa. But until the constructal law we didn't know why.

As the sun beats down, it heats the water on the surface. Warmer water is lighter than colder water. The warm water moves horizontally to climb on the cold water, while the cold moves in the opposite direction to slip under the warm. The upper layer has horizontal momentum (literally movement) and the lower layer does not. The upper layer reaches down, grabs the cold water, and drags it along with it. The fast surface water entrains, or literally pulls along, the slow below-surface water so that the two tend to flow at about the same speed. In physics this entrainment phenomenon is described as a process of transferring motion (momentum) from the fast to the stagnant to achieve a sort of equilibrium. This occurs because the tendency in nature is to equilibrate not only the hot with the cold but also the slow with the fast—equilibrium means uniformity in every respect.

Though the naked eye might see that water is flowing, what's really going on is the flow of momentum (motion), from fluid packets that move quickly to packets that move slowly. The question of design in nature emerges when we ask how this momentum is transferred. There are two design choices. Laminar flow consists of smooth parallel blades of fluid (one blade rubbing and sliding on its neighbor) as they all move forward. Turbulent flow is characterized by whirling eddies (chaotically) moving forward and perpendicularly (that is, rolling). In both laminar and turbulent flows, the momentum is transmitted *vertically* downward, that is, perpendicularly to the horizontal movement of the water.

You can observe this same phenomenon by looking at a plume of smoke rising from the smokestack of a power plant; here the fast fluid (the gas) moves vertically as the smoke shoots up in a slender, cone-shaped column. As it transfers its momentum perpendicularly, the plume expands horizontally to encompass more and more air around it. The plume becomes wider as it lifts more and more air with it. The same thing happens when you light a cigarette or a candle. Initially the smoke, which is warmer and lighter than the surrounding air, rises as a slow straight column. As it picks up speed, the flow changes; it starts to create rolling whirls called eddies (or *tourbillons* in French, from which we derive the words "turbulence" and "turbine"). The cigarette smoke is the upside-down drawing of the airstream entrained downward by the falling toilet paper illustrated in Figure 5.

What the constructal law reveals is that flows choose the design that will better facilitate their movements at any given time. The constructal law also allows us to predict the point at which this transition between laminar and turbulent flow should occur to facilitate the flow of momentum.

We find that all fluids flow in this sheetlike motion, called laminar flow, when this is the better way to spread momentum. This is the flow design when the flow stream is narrow and slow enough. But when the stream is sufficiently thick and fast, the design changes to turbulent flow.

Throughout this book, I refer to flows that are "slow enough" or "fast and thick enough." This language is necessarily vague because it covers a broad array of phenomena moving in diverse environments. (It is also why I say a particular design is "good" for flow.) Each particular flow is unique. What is fast and thick for one type of flow is slow enough for another. The universal principle is that when the specific threshold is reached in every case, the flow changes (it clicks!) into the better design.

Nevertheless, as I will now show, I have used the constructal law to predict when that transition should occur in an imaginary flow. I have tried to minimize my use of math in this book, but it is important here, and a few other places, to show you the work behind my conclusions. The natural tendency of selecting the

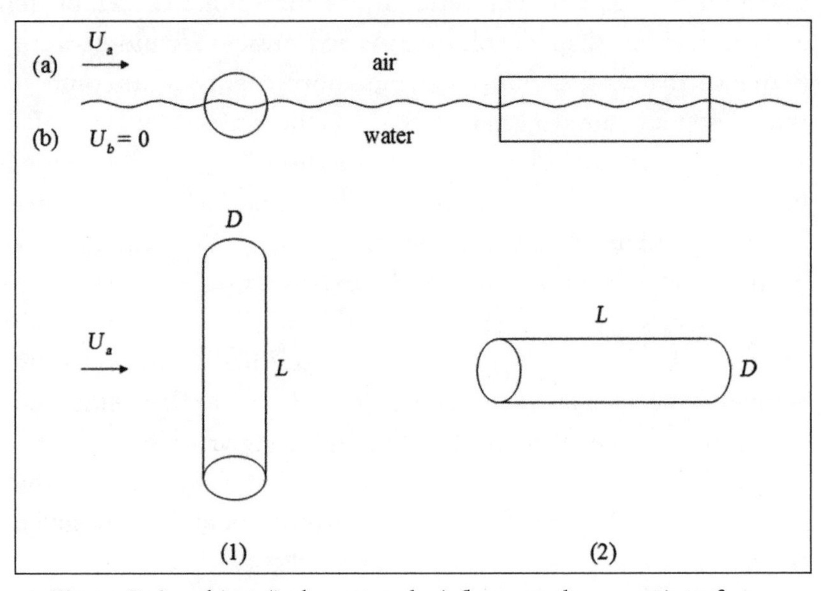

Figure 9. An object (iceberg, tree log) floats on the ocean interface between two fluid masses (a), (b) with relative motion. The atmosphere (a) moves with the wind speed U_a, while the ocean water (b) is stationary. Momentum flows downward from (a) to (b). The constructal law calls for the generation of flow configuration such that momentum flows more easily. The floating object is the mechanism by which the atmosphere (a) transfers momentum to the ocean water (b). The extreme positions of this mechanism are (1) and (2). The forces with which (a) pushes (b) via the object are F_1 and, respectively, F_2. The selected configuration must be (1), because one can easily show that F_1 is greater than F_2 when the cylinder's length (L) is greater than its diameter (D). Momentum flows from (a) to (b) at a higher rate through configuration (1) than through configuration (2). This is why icebergs, waves, and debris orient themselves perpendicularly to the wind direction. This prediction is confirmed by all the forms that drift sideways (perpendicular to the wind) on the ocean: icebergs, debris, water waves, abandoned ships, etc. The emergence of the turbulent eddy is the same design phenomenon as the selection of configuration (1) for the floating object.

flow configuration so that momentum flows more easily across the flow is illustrated in Figure 9.

Figure 10 shows the shear flow between the fast and slow regions of the same fluid, (a). The threshold or the point at which the flow switches from laminar to turbulent occurs at each flow system's Reynolds number. When the shear flow is thin and slow such that the Reynolds number is less than 10^2, the more effective design is viscous shearing (laminar flow). When the shear flow is thick and fast such that the Reynolds number is greater than 10^2, the more effective design is eddy formation (turbulence).

The Reynolds number 10^2 marks the birth of the first, smallest eddy. The theoretical leap is that the constructal law demands the occurrence of eddies. In constructal theory, the eddy is predicted, not assumed, not seen and then described, not the aftermath of an assumed disturbance. Every eddy is generated at the intersection of the two curves in the lower part of Figure 10 and expresses the balance between two momentum transport mechanisms. Every eddy is a package of two flow mechanisms: streams (the roll) and viscous diffusion (laminar, inside and outside the roll).

In summary, when the flow is fast enough, the turbulent flow becomes the more effective way to transfer momentum laterally. The key point is that the water (like the smoke from a chimney or a cigarette) has two design options: laminar flow or turbulent flow. When laminar flow entrains more of the surrounding fluid, that is what we observe. That happens when the flow is a trickle, narrow and slow. When the flow is wider and faster, turbulent flow (rolling eddies) rules the momentum transfer process. Like the curls of cigarette smoke, the oceanic eddies are spinning wheels that rearrange the water so that it can transfer momentum more easily and mix the ocean more effectively.

Why do isolated systems evolve toward equilibrium? We don't know. They just do. This is why the second law of thermodynamics is called a first principle—because it cannot be deduced

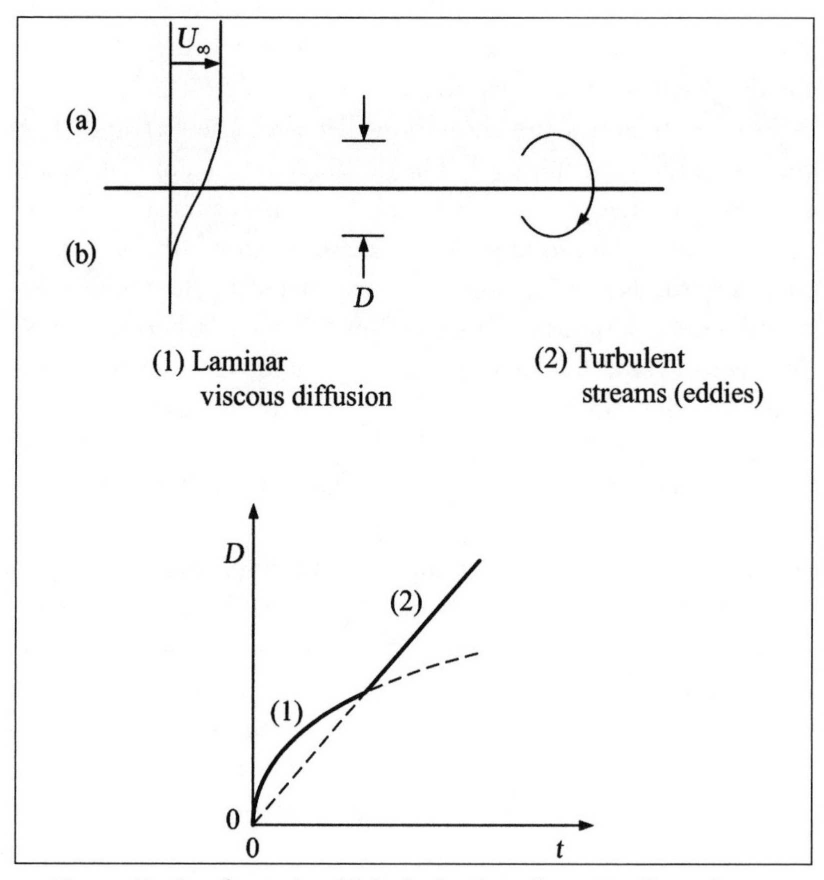

Figure 10. Configuration (1) is the laminar flow. Configuration (2) is the eddy flow, or the wrinkling, rolling, and thickening of the shear layer. The lower part of the figure shows how the shear layer thickness D grows in time, that is, how momentum is being transferred from the fast to the slow. In the laminar configuration (1), momentum is transferred by viscous diffusion, and consequently D increases as $t^{1/2}$, where t is time. In the eddy configuration (2), D increases in proportion with t, because D scales as $U_\infty t$, where U_∞ is the speed of the moving fluid. The two $D(t)$ curves intersect when the Reynolds number based on D and U_∞ reaches this threshold

$$\mathrm{Re} = \frac{U_\infty D}{\nu} \sim 10^2$$

where ν is the kinematic viscosity of the fluid. The constructal law calls for the configuration that generates the larger D (more momentum transferred) at any point in time.

from other principles. The same is true for the constructal law. Where the second law describes the universal tendency to flow from high to low, the constructal law describes the universal tendency to generate evolving configurations that facilitate that flow. The second law and the constructal law, then, are two different first principles. Together they capture nature much more firmly than the second law alone.

To see the constructal law at work, imagine a box filled with a compressible fluid (air), with parts of high pressure and low pressure. The high pressure will move air mass toward the low pressure. When I was a student at MIT, my professors and I never questioned the drawing involved, the path of that movement. I am now amused to see that in my first book, *Entropy Generation Through Heat and Fluid Flow* (1982), I drew the diagram of a flow system shaped as a blob, as in Figure 8; like everyone else, I took the drawing for granted. On one level, this is akin to my saying "I traveled from Paris to Rome last night" without any consideration about how I got there (bus, train, plane), the route I took, or the time it took. In fact, the journey is everything.

Through the years, doing my best to serve (to protect, to defend, and to save!) the discipline of thermodynamics and apply it correctly, I recognized the necessity and centrality of design. I realized that that black box was not just an abstraction for input-output analysis. It was not filled with scientific abstractions but with drawings that were moving and morphing.

In retrospect it seems surprising that scientists had long ignored this basic phenomenon of design generation and evolution. After all, the concept of flow design was crucial to the work of Carnot and all who have followed him. Nevertheless, they did. As John Steinbeck noted in his wonderful nonfiction book on marine biology, *The Log from the Sea of Cortez*: "Often a very obvious thing may lie unnoticed."

Let's return to the box filled with regions of high and low pressure. If nothing interacts with the box, the second law says that over time the high and low regions will equilibrate themselves. It says nothing about flow, configuration, evolution, and design. In the very beginning the geometry of the flow system is missing—because no flow has occurred. The constructal law accounts for a different phenomenon: the generation of flow configuration in time. This other phenomenon facilitates the access of the high-to-low currents through the box.

Geometric form is generated in natural systems that are internally "alive" with flows and driving gradients (for example, temperature and pressure). Such systems are not in equilibrium internally. They are not dead; they are on the move. In order to get from here to there, everything must create a path. People, then, are only half right when they say things *seek* the path of least resistance. Instead of finding these already cleared paths, flow systems *construct* their own flow architectures and body rhythms that enable them to move more easily.

As we look around us we see striking similarities in the architecture and evolution of simple and complex flow systems: lung structure, river basins and deltas, animal movement, respiration, solidification, etc. Consider the blood vessels in your body. Their cross sections are nearly round because, in the abstract, the mathematically circular cross section is the design that offers the least resistance to the mass that flows though it. The cross sections of animal flow channels are never perfectly round because the animal body moves and morphs. Nearly round, however, performs almost as well as the mathematically round (think about it as you press the veins visible on your forearm: you do not faint, that is, you do not feel the effect of having made the cross sections of those blood vessels imperfect). Likewise, moles and earthworms create nearly round tunnels when they burrow through the ground. Same goes for civil engineers whose tunnels and mine shafts are nearly round—for that matter, so, too, do naturally formed tunnels such as the underground caves, the shafts of volcanoes, and those formed by the jets of fluid that pierce water and air. Using the constructal law, we can predict that when one thing goes through another, we will find a rounded cross section.

Coincidences that occur in the billions are a loud hint that a universal phenomenon is at work. It suggests that there is a single principle of physics—not of biology, or geology, or sociology, but of everything—from which the phenomenon of configuration and rhythm can be deduced without any recourse to empiricism.

The movement toward equilibrium—and the pull of gravity—puts things in motion. The constructal law proclaims that nature will generate configurations to facilitate this flow and that there is a direction in time to this phenomenon of configuration generation. Nature is indifferent, impersonal. But it does have a tendency—to mix and to move everything on Earth, more mass, moved farther. This constructal tendency fuels all evolution and design.

Through this discovery, we turn the page on the long-raging debates about direction, purpose, evolution, and design. We begin writing an entirely new chapter in the book of science that shows how the complex forms of shape and structure we see all around us arise from the laws of nature.

CHAPTER 2

The Birth of Design

Before I was a scientist I was an artist. Okay, maybe that's going a little far, but as a child I loved to draw. I always had a pencil in my hand, drawing racing motorcycles, horses, and everything else that was on the move (hint: The constructal law started then). Seeing something in these designs, my parents sent me to art classes after school. In my office at Duke, I still have a drawing of two Dutch ships I made in the fourth grade—and a portrait of my younger daughter, Teresa, I drew in a restaurant in Rome in 1990.

In retrospect, I realize that I had the feeling that I "saw" how things moved and how they fit together. Drawings provide the first clue about operation; they begin to tell us what something is by suggesting what it does. As Michelangelo reportedly observed five centuries ago, "Design (drawing) . . . is the root of all sciences." I am always offering my amateur sketches to my students at Duke so they can see what we're talking about, to remind them that the physical world is not made up of ivory-tower abstractions but images with shape and structure that move across the landscape and conform to physical laws.

Where a drawing provides the outline of an object, science allows us to burrow in and see how it functions—drawings show us the parts; science shows us how they move. Science is effective because it is concise. It converts physical phenomena into statements, formulas, and mathematical equations that have great explanatory power. In the process, it also tends to sever objects from their natural state. The mighty Danube ferrying water from central Europe or an elegant antelope jumping across the savanna loses its essential character when translated into data.

Speaking practically, this wouldn't matter if design were simply an aesthetic concern. Science goes with what works, pleasantries be damned! The rendering of nature as charts, tables of numbers, graphs, and equations has opened up vast areas of knowledge and understanding. It underpins much of my work. However, it has also blinded us to deeper truths. Like the muckraker in John Bunyan's classic novel *The Pilgrim's Progress*, it has focused the researcher's gaze downward on his own small patch of ground.

When we raise our eyes and look around, we encounter a wondrous world of living drawings: birds and airplanes painted against an azure sky, pine trees and skyscrapers reaching for the heavens, rivers and roads snaking across the Earth's surface. If we take a closer and wider look at the same time, we also see how much these images have in common: similarities in shape and structure so numerous that they can't be the result of accident.

The constructal law makes "design" a concept in science. It reveals that scientists have been digging in the wrong patch when they ignored configuration or simply took it for granted. Design is, in fact, a spontaneously arising and evolving *phenomenon* in nature. Design happens all the time everywhere, not as the result of one mechanism but as the expression of a law of physics like Galileo's principle of gravitational fall and the laws of thermodynamics.

Language can make this hard to grasp. The constructal law uses "design" as a noun that describes a configuration, which is known by many other names: image, pattern, rhythm, drawing, motif, etc. This sense, however, has been conflated with the verb "to design" that refers to the power of the human brain to contrive and to project images and linkages to new, higher planes. To design is human. It is human to absorb images that invade us, to reflect upon them in our minds, and to use them as personal catapults to make our drawings and devices so that we become a better and better species moving more easily on the landscape. In fact, we are so tied to the technologies that enhance our movement that we have evolved into a *human-and-machine* species (more on this later).

The verb "to design" has been monumentally unproductive in our quest to understand design in nature for three main reasons. First, it led to the common view that the things humans design are "artificial," in contrast to the "natural" designs that surround us. This is wrong, because we are part of nature and our designs are governed by the same principle as everything else, the constructal law. Second, it has led some of us to search for "the designer"—God, or an individual, who must be behind every design. Science is not and never was the search for "the designer." The name for that much older search is religion. Finally, it has led other, more scientifically minded people to reject the idea of design in nature as part of a broader repudiation of the traditional idea of a designer.

The constructal law tells us to stop looking for a phantom designer—there is no single mechanism or design-generating entity that can be found in river basins, blood vessels, transportation systems, etc. It teaches us, instead, that design is a phenomenon that emerges naturally as patterns. It also tells us that this evolving shape and structure is predictable. That is, if we know what is moving through a flow system, we can predict the sequence of designs that will emerge and evolve to facilitate the currents that run through it.

This starts with a drawing, or to use a better metaphor, the first frame of a movie—with what something looks like at a given moment. But nature does not exist in freeze-frame; it is dynamic, ever evolving. As the film rolls, the drawing changes over time in one direction: to flow more easily. I'm tempted to give this never-ending movie a grand and catchy title like *Gone with the Flow* or *I, Constructal.* This thrilling blockbuster details how flow systems configure and reconfigure themselves to overcome the friction and other forms of resistance that hinder them. Faster, easier, cheaper in terms of fuel (useful energy, exergy) used and materials required for movement: that is the flow system's mantra.

In this chapter we show that evolving design in accordance with the constructal law is a universal phenomenon by focusing on three flow systems that would seem to have little in common. The first comprises the man-made cooling systems designed to remove heat from electronic devices. The second is the river basin that represents inanimate, nonbiological systems. The third is the system of blood vessels that carry oxygen and energy throughout our bodies. Each of these systems has been explored in great depth through the years; we know a tremendous amount about their shape and structure. But the systems have also been studied in isolation. This approach has led researchers to consider them not just apples and oranges, but apples and sports cars, oranges and shoes. The constructal law reveals that these flow systems generate strikingly similar designs in order to facilitate their own movement.

All three examples have at least two things in common. First, they are steady-state systems, that is, the currents that run through them (heat, water, blood) do not change much. Second, all three systems face one of the most common challenges in nature: how to move currents (of heat, fluid, people, goods, it doesn't matter) from a point to an area or from an area to a point. This may sound like an abstract idea, but it is one that affects all of us every day. The movement of water from the reservoir (a point) to the various faucets and taps throughout our community (the area) is one example. So is the movement of sewage from each home and business (the area) to the treatment plant (a point). When we leave our homes each morning to go to work or the mall or to take the kids to school we become part of the volume of people flowing from the area (our neighborhoods) to various points within our local communities. We travel along networks of roads designed to get us where we want to go in the most efficient manner: faster, easier, cheaper. When we zoom along to our destination, obstacles have been mitigated if not eliminated. When we are stuck in traffic because of bottlenecks, we pay the price for outdated design.

I faced an area-to-point problem earlier in my career when I was designing cooling systems for electronics. My objective was to install as much circuitry as possible into the fixed space of a machine. Like everything else that moves, electronic components generate heat as they function. The heat is the result of dissipating (destroying) in the electrical resistances of the circuitry the electrical work taken from the wall outlet, in order to push all the electric currents through the circuitry. The more electronics you squeeze into a confined space, the hotter it gets.

The modern world of multifunctional cell phones and laptop computers hundreds of times more powerful than the room-size units that represented cutting-edge technology during the 1950s would not be possible if engineers hadn't figured out how to channel away the heat, making these devices smaller, cheaper, and faster all at once. The burgeoning era of nanotechnology—which promises machines smaller than an eyelash—depends in great part on our ability to make those tiny workhorses operate without melting. Most people don't give it much thought, but countless things we take for granted depend on our ability to remove heat.

There are many ways to cool a system. You can blow air on it, as a fan does inside your personal computer, or you can run coolant through it, as the tubes of Freon in many refrigerators and air conditioners do. Both approaches are effective, but they involve various costs—just as there's no free lunch, there's no free cooling. Blowers and cooling tubes take up lots of space. This doesn't matter much with large appliances. But when we are measuring things in micrometers (a millionth of a meter), we need a better way.

My challenge while doing theoretical research in the early

1990s was to find a way to cool a solid block of circuits so small that it had no space for coolant coils or air. I had to find a way to cool, without a moving fluid, the inside of an electronic rock that was constantly generating intense heat.

I began with pencil and paper. I drew a rectangle filled with circuits that produced heat at a fixed rate. This heat was the moving current in the system. My goal was to create a flow design that would pull the heat most efficiently from the *entire area* of the rectangle. After all, the heat does not stay within the circuit but travels to its environment. Remember, flow occurs in the physical world; thus, it is always tied to space, to geography. It is always mixing with and churning all around it, as we saw in chapter 1 when the warmer and faster-moving water on the ocean surface entrains the colder, slower-moving water below it. In my work, the best way to move the heat current was through solid-body conduction—by channeling the heat out of the core, solid region closer to the perimeter. I made two key decisions. First, I posited that the circuits are mounted on a structural material that is a relatively poor conductor of heat. Then, I imagined that I placed a slender strip of highly heat-conducting material—such as graphite, gold, or diamond—down the center of the rectangle (A_0 in Figure 11). This would channel the heat from the surrounding area. There were no moving parts, so how would the heat leave the system? High temperatures drive heat currents to low temperatures. This natural flow would lead the heat to the cooler space first inside, near the perimeter, and then outside the system.

Through this design, I hit upon the seminal ideas that govern the design of everything that moves and flows, ideas that I would express a few years later as the constructal law. The first breakthrough arose from my decision to use two kinds of material—pieces of low and high conductivity. This choice proved to be a stroke of luck, because it encompassed the two major elements that cover the entire area of every flow structure: the channels and the finite-size spaces between adjacent channels (called "interstices"). In all designs, currents move slowly over

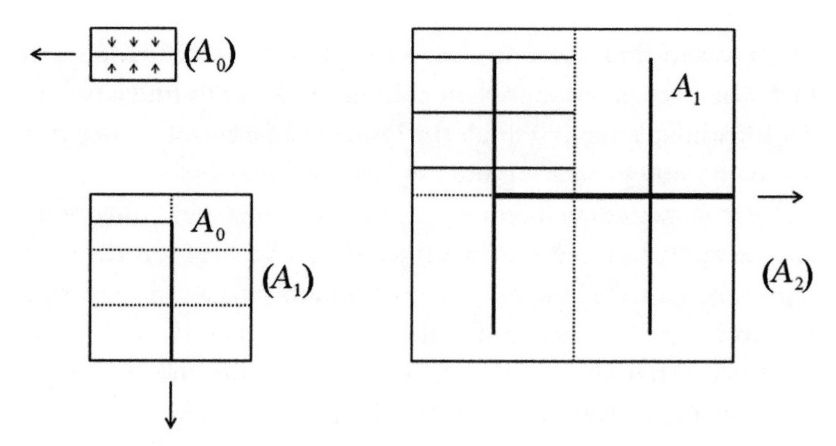

Figure 11. How to bathe an area with area-to-point flow. Start with an elemental system (A_0) and shape it such that the resistance along the channel (the black centerline) is matched by the resistance above and below it. Next, the architecture of the first construct (A_1) is discovered in the same fashion. Note that A_1 is an assembly of elemental systems centered on a main channel (flowing downward), while the nerves of the elemental systems serve as tributaries to the main channel. The second construct (A_2) is an assembly of first constructs (A_1), and its architecture is discovered along the same mental route as from many A_0s to a single A_1.

relatively short distances through the interstices, and faster over longer distances through the channels. Because the constructal law is about pulsing, morphing things on the move, the channels and interstices are not static objects. They are not paintings or rocks that just sit there. They are designs that emerge and evolve to facilitate the flowing of the whole. It takes two to tango and this is the dance of flow design.

To see how, let's return to the circuits shown in Figure 11. The heat they generate is diffuse and disorganized. It moves slowly over the relatively short distance of the interstice to the highly conductive center strip. We cannot see diffusion. Once the heat is absorbed into this central channel, its flow becomes organized and it moves quickly down the channel and out of the system. We can see the channel.

Now consider rain that falls on a hillside and seeps into the ground. Like the heat given off by the circuits, its initial flow is diffuse, disorganized—the water seeping into the ground encounters a large resistivity as it moves through the porous soil (the interstices). Eventually the seeping water coalesces to form a rivulet, the first tiny channel. It does this by displacing the existing soil grains and connecting the pores between them. Because the rivulets evolve from what had been in place before they arose, the tiniest rivulet has the same thickness as the pore and the grain. Here's the key point: By generating this design, its flow becomes organized and visible; it moves faster, more easily toward the quicker streams and eventually the running river.

We see the same thing in our circulatory system. We start in the heart, which pumps the blood through the aorta that branches off into ever-smaller channels (arteries) and very narrow rivulets (capillaries) that spread oxygen and useful energy to every cell (the interstice) through diffusion. The tiniest capillary has the same thickness as the "grain" that existed before it, the cell. Similarly, the system that carries the blood back to the heart goes from the small capillaries to larger veins and finally to the large vena cava, where the blood will begin its journey once again. In this case, too, design emerges because it facilitates flow.

The efficiency of this design is clear when we see how this architecture guides our morning commute. It usually begins with a walk to our cars, which we drive down small streets; just as capillaries are necessary for blood to reach every cell, local roads are needed to reach everyone in the community. If we had only local roads, it would take a great deal of time and fuel to travel long distances. So we construct highways into which the local roads flow. Once we leave the highway, we again travel on local roads to get where we're going.

Highways offer obvious benefits, but this does not mean that they are the best option everywhere, because they cannot accommodate every commuter on the area they serve—just as the Danube cannot reach every seeping hillside on the continent and the aorta cannot move blood to every cell in our bodies. The slower, shorter paths are better for this. Thus, the slow and short flow works with the fast and long flow to move currents efficiently.

What flows through the system—heat, water, blood, people—is not nearly as important as the fact that every tree-shaped flow architecture is defined by these two flow regimes. It is this phenomenon that leads to design in nature. To see how, let's imagine once again rain falling on a hillside. Initially the rain covers the ground evenly; when the rainwater moves, it flows like a sheet over the entire hill. How would you draw the movement of this water? You couldn't, because it has no discernible pattern. Design starts to appear when the rainwater on the ground coalesces to form a rivulet. It is this transition to a new way of flowing that provides the *contrast*, which is the essence of design. This transition is the birth of design.

To think about water flow, take a white sheet of paper and imagine that it is a picture of wet ground. As the rain hits the ground and seeps into or flows evenly over it, nothing changes. To have something to draw, something must happen.

Draw a short line in black ink on the paper. This is the first rivulet formed by the water. Now we have black *and* white: channels and interstices. We have pattern. We have not only drawn the first rivulet but also have given structure to the white space that represents the rain flowing over the ground. As you add more black lines, representing the streams and rivers that evolve from the rivulets, your drawing becomes more complex.

The necessity and interdependence of channels and interstices is a point that was not fully appreciated before the constructal law. Design emerges over the *entire area*: It is the white space of oozing water that sustains and nourishes the black lines of the evolving channels that move that water more easily. It is important to remember this as we consider other flow systems. When we look at the circulatory system, for example, the eye focuses on the intricate, treelike structure as the aorta branches off into arteries and capillaries. But we can recognize this design (the channels) only because it stands in contrast to the surrounding tissue (the interstices) that is being fed.

This phenomenon also tells us something else. In order to move more easily, the flow system acquires geometry, design. It creates a path, many paths, connected in certain ways. Before the constructal law, we could see that rainwater coalesces to form rivulets, but we did not know why.

The constructal law accounts not only for the emergence of design but also its evolution. To see how, let's return to the cooling system I was designing in the early 1990s. I had great freedom; I could design the path, write and direct the movie of morphing flow. The question we can ask now is: Would the designs I made when every solution was possible, using mathematics and engineering, resemble those that arise spontaneously in nature?

I began with the fact that all my circuits were giving off heat. If I left things as is, the heat would move out of the system too slowly, leading to a meltdown. So I placed that strip of high-conductivity material down the middle, to create a better, faster flow. I, of course, had an objective when I did this. The rainwater, by contrast, has no mind of its own. But two natural phenomena are part of its being. The first is gravity, which pulls the water to lower ground. The second is the constructal law, which accounts for the fact that rain will form the first rivulet when the flow of water becomes large and rapid enough. That is, when the flow is slow and short, diffusion is the way to go. But when the flow intensifies, an organized structure with streams and channels is better.

I called my first drawing the *elemental construct*. Then I got more ambitious. I thought of cooling a larger area. My objective remained the same: to facilitate the flow of heat toward the sink so that the entire area was cooled efficiently. As I added circuits, I increased the amount of heat in the system. A single strip of high-conductivity material, or cooling blade, would no longer be sufficient to handle the extra load—just as a small road cannot handle all the cars in a city. In simple terms, the heat would back up, riddling the area with congestion, causing overheating.

One possible solution was to place a cooling blade next to each circuit. This would have cooled the system, but it also would have made a heavy machine. Efficiency, including lightness, is a hallmark of good design. So I changed the positions and thicknesses of my blades and of the adjacent interstices to put the right-size blades in the right places.

I did not know it at the time, but this is the same thing that happens in a river basin. Even as my circuits were bombarding my rectangle with heat, so, too, does rain cover a hillside with water. In time the right-size channels emerge to handle the flow over the entire area efficiently. Just as I did not place a cooling blade next to every circuit, nature does not place a main river channel next to every hillside. Instead, we find a hierarchical pattern of many small rivulets, streams, etc. and a few large channels. And, while all river basins have the same basic design, they are all different in that they have evolved their own combination of small and large channels to serve the particularities of their location.

This intricate design does not evolve at once. At first, seeping is a good way to flow. The first rivulet forms when resistance builds to the point—which we can predict mathematically if we understand the environment in which it is emerging—that this becomes an easier way to flow. This is the elemental construct. As the volume of water increases, the rivulet coalesces with other rivulets, making a larger channel. This is the *first construct.* The process continues, creating a series of larger and larger constructs until a river basin emerges that serves the flow of water over the entire area through the right balance of multiscale channels.

Although these various-sized channels emerge in response to the specific resistance encountered by the flowing water, the overall balance of the system is achieved through a universal design balance we will explore in greater detail in chapter 7. It is this: The resistance to moving slow and short should be comparable with the resistance to moving fast and long. That is, when we look at the evolving river basin, we find that the time water spends seeping through the ground (moving slow and short) should be roughly equivalent to the time it spends flowing long and fast in the channels.

I used pen and paper, and later computers, to design and test the performance of various ways to cool the entire area. One solution was to place a strip of high-conductivity material down the center of each rectangle, producing a design of parallel lines. This proved to be less efficient than having a smaller branch run off the single main channel at a slightly less than 90-degree angle. This was my first construct (see Figure 11, (A_1)). To dissipate the heat generated in areas even farther away, it was best to shoot other branches off these. When I looked at the drawing, I saw that I had created a treelike pattern. It did not occur by accident. It was completely deterministic. That was why, a few months later in France, I knew Prigogine was wrong when he proclaimed that the treelike patterns we see in nature are the result of chance.

It didn't strike me until much later—after I had discovered the constructal law—that as I did this work I was engaged in a version of that old riddle: Which came first, the chicken or the egg? As I sat at my drawing board, I was essentially playing the role of the constructal law for my rectangles of heat flows: I was the clock and the evolutionary mechanism morphing the design for better flow. Just as in nature, I did this by changing the geometry freely: the drawing of the streams and channels. That is, I improved the performance of the whole system by changing its design. When resistance to heat flow by conduction built up in one channel, I reduced it by adding a branch or by bringing branches together. I could not eliminate any of the resistances. Imperfection is an inevitable phenomenon and a necessary part of design. As we will see in more detail throughout this book, good design involves *the nearly uniform distribution of imperfection* throughout the entire flow system.

Here's another key point: I was not trying to mimic nature. I was not even looking at nature. I was living in my mind, working from pure theory of how things *should be*—better flow through the reduction of global imperfection. My approach corresponded with the one in nature, because this is the way to provide access to flow. The natural way is "anything goes." Call that the egg part. The chicken comes in because I realized my efforts stemmed from the constructal tendency of everything to find better ways to flow. Thus, even as I was acting as the constructal law, I was governed by it. My "artificial" designs were completely natural.

Four subsequent experiments illustrate this. In the first, Sylvie Lorente of the Institut National des Sciences Appliquées de Toulouse, Wishsanuruk Wechsatol of King Mongkut's University of Technology Thonburi in Bangkok, and I asked a computer to move a steady stream of water with less and less resistance out of the center of a disk to six users equidistant from one another

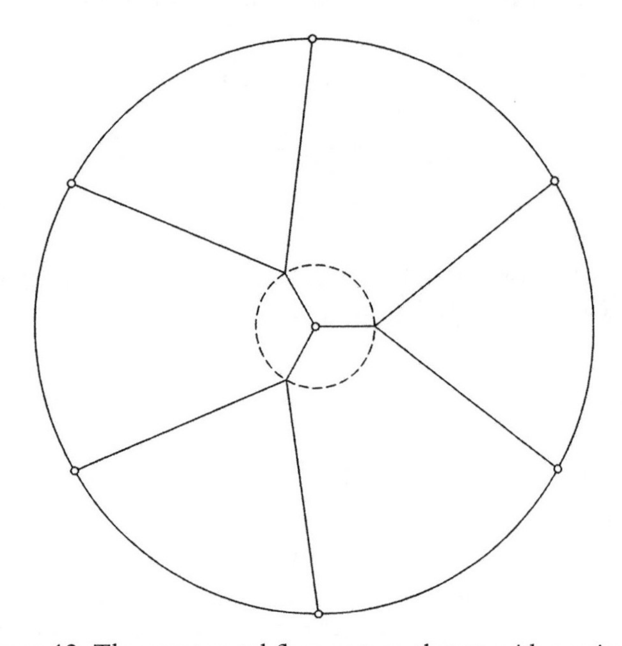

Figure 12. The constructal flow pattern that provides easiest access for laminar flow between the center and six equidistant points on the circle. The bifurcations are located on the concentric circle indicated with a dashed line.

on the perimeter. We gave the computer total freedom to solve this problem—it could use any material of any shape and size. Figure 12 illustrates that it generated a treelike structure.

Then we ramped it up, asking the computer to serve twelve and then twenty-four users on the perimeter—that is, to morph the design to serve a larger area. As we see in Figure 13, it once again offered more elaborate treelike patterns, which are actually quite regular and simple, that is, easy to remember.

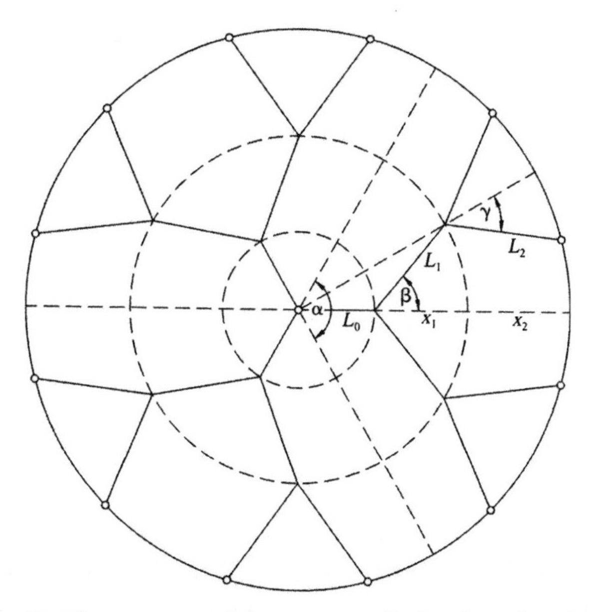

Figure 13. The constructal flow pattern for laminar flow between the center and twelve points on the circle. Note the two levels of bifurcation.

In the second experiment, J. D. Chen of the oilfield services company Schlumberger-Doll Research took two thin pieces of glass roughened with evenly spaced dimples. He covered one side of each piece with glycerin, a viscous fluid. He sandwiched them by putting the roughened sides together and laid them down on a table. Then he took a syringe filled with dyed water that he injected through a central hole in the top plate. The dyed water displaced the glycerin and created the now-familiar treelike pattern predicted by our computer simulation and pencil-and-paper calculations. That is, the water flow organized itself into a predictable pattern that evolved into a treelike structure in order to flow more easily (Figure 14). The less viscous fluid (water) flowed through the channels, while the more viscous one (glycerin) flowed through the interstices, not the other way around. The river basin cannot be a tree-shaped structure of wet mud surrounded by flowing lakes of water.

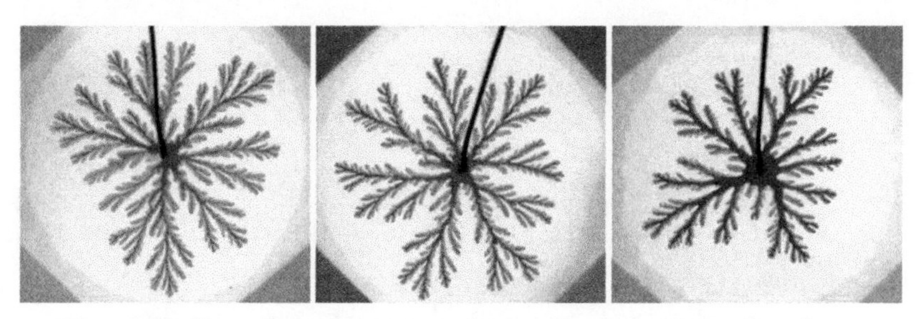

Figure 14. Tree-shaped patterns obtained by injecting colored water into a layer of glycerin between two glass plates.

The third experiment was one that you can easily duplicate at home. Grind some coffee beans finely and put them in a pot. Add water and bring it to a boil. You want the finest sediment that is still emulsified in the liquid, not the coarse grounds that settle in the bottom of the pot. Take the pot off the heat, wait three minutes, then carefully pour off the liquid. Wait another 30 minutes to allow more of the grounds to settle to the bottom. After that, pour out almost all the liquid, so that the remaining sediment has the consistency of soft honey or paint. Use this liquid to wet the entire area of a concave surface, such as the inside of a funnel (see Figure 15).

At first it doesn't seem like anything is happening. In fact, the water is flowing volumetrically (diffusely) over the entire area and it is encountering a great deal of resistance. Then patterns start to emerge as the seeping water coalesces to form rivulets and then branches as it organizes itself into a predictable geometric pattern—

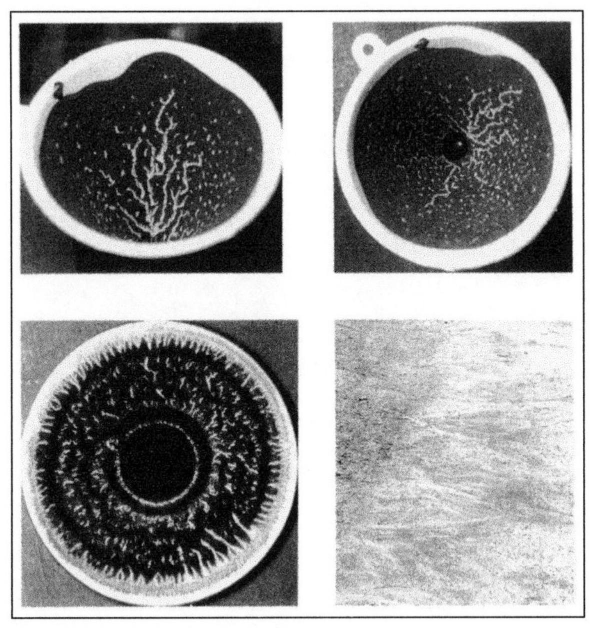

Figure 15. The formation of the smallest rivers in the drainage basin of a funnel coated with unfiltered coffee sediment. The funnel was held vertically upward, and the photograph was taken at an angle and from above. Note the marriage of shapeless flow (disorganization, diffusion) and flow with tree shape and structure (organization, streams) at the smallest, finite scale. Trees form all around the funnel and are visible from above. *Bottom right:* the first rivulets after the rain, on a sloped, sandy terrain.

an evolving treelike design—to reduce resistance and flow more easily as a whole. You can actually see how the water pushes the coffee grounds out of the way to form the first rivulets.

Finally, to show how river basins evolve in time, several of us made "movies" documenting how they generate their area-to-point tree-shaped designs in accordance with the constructal law. In the first, Stanley A. Schumm of Colorado State University and his student R. S. Parker covered a 15-by-9-meter area with sand in a laboratory. Then they pelted it with a steady and uniform artificial rain. To mimic a natural hillside, the surface was flat and tilted slightly so that the water would drain to one

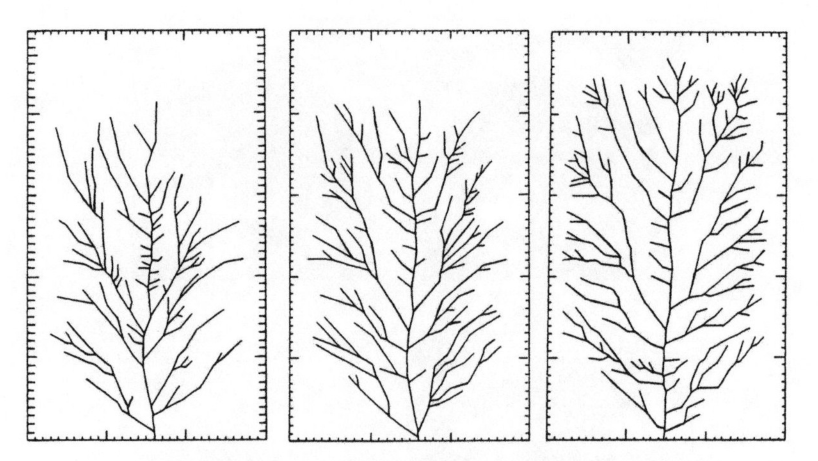

Figure 16. The evolution of an artificial river basin under uniform rain on the floor of the laboratory.

side. The rain wet the sand uniformly, yet channels developed nonuniformly, in treelike fashion. Their development never ends. The tree shape keeps changing so that the collected water gets out of the area more and more easily (Figure 16). This is evolution, reproduced in the laboratory.

The second movie was a computer simulation based on the same scenario of steady uniform rain on a square territory, in a model of soil erosion in which grains are removed from spots where the local seepage velocity is high enough. The dislodged grains created channels with markedly higher permeability, and the channels formed a tree shape that grew into better and better tree shapes for moving more water more easily. The soil is uniform, with the same erosion characteristics everywhere. The velocity threshold needed to remove one grain was the same over the territory. This is why the flow structure that emerges is a symmetric tree (Figure 17).

The third movie was generated on the computer with the same erosion model, except that the soil structure and properties were not uniform. This time, the erosion threshold velocity had values distributed randomly over the square basin. The resulting shapes had a random appearance (Figure 18). This is due to the

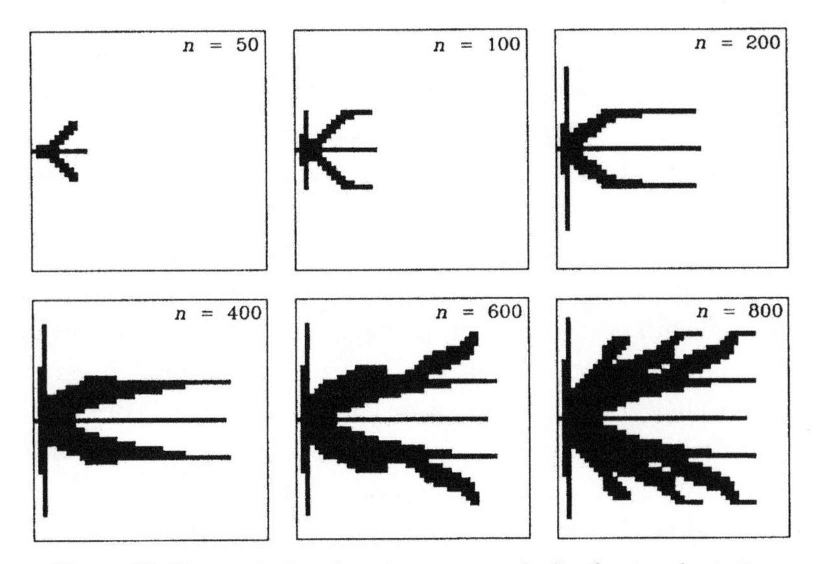

Figure 17. The evolution (persistence, survival) of a river basin in a porous layer with uniform resistance to erosion (n is the number of grains removed by the erosion process, that is, the measure of time).

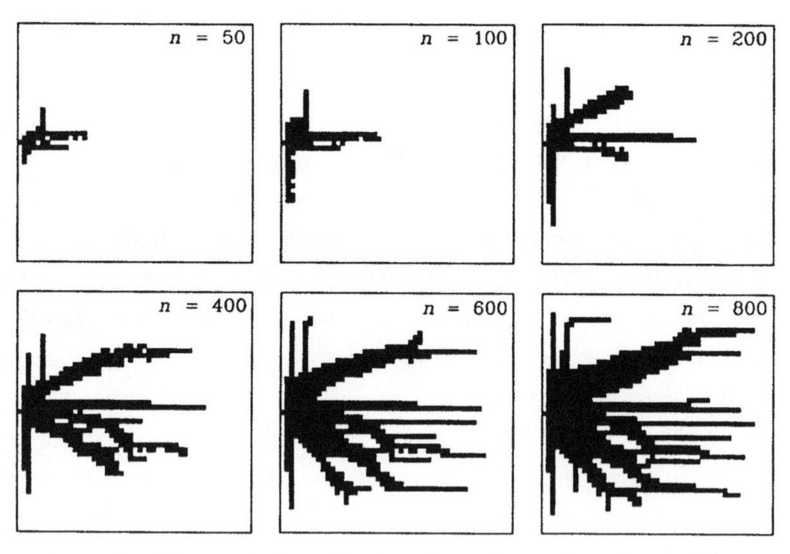

Figure 18. The evolution of a river basin in a porous layer with unknown (random) distribution of resistance to erosion. Compare this to Figure 17.

assumed random geological properties of the terrain, not to a presumably random tendency of generating configuration. This is the origin of the erroneous view that tree-shaped river designs are the result of chance. The geology and the conditions are, but the constructal law is not acting capriciously.

The principle that generated all these configurations was the same and it was deterministic. The second and third movies were reproduced in every frame of the movie, when the erosion simulation process was repeated on the same basins, with the same geological erosion characteristics. These experiments confirmed the fact that natural river basins of all sizes acquire *evolving design*. Lightning bolts; trees; the air passages in the lungs; the arteries, veins, and capillaries in the circulatory system; and every other point-to-area or area-to-point flow in nature always generate predictable patterns in order to flow more easily.

The treelike pattern isn't the only configuration we see in such systems. River basins and blood vessels exhibit other design features that reflect their tendency to seek greater flow access. Several of these are well known, but until the constructal law we did not have a single explanation that predicts and unites these phenomena.

River basins reflect firmly established rules that relate their channel numbers and lengths. These include the fact that rivers meander—that is, they have a snakelike pattern whose wavelength is proportional to the width of the channel. In addition, the width of the channel is about five to ten times greater than its depth. For example, standing at the widest point of the Danube River, which is about one kilometer, we know the water there is quite deep. By contrast, we know that the rivulet that forms under a faucet in the garden is not deep because it is not wide.

Other scaling patterns—known as the rules of Horton, Melton, and Hack—have been known since the 1930s and are based on extensive measurements of river basin geometries all over the world. As we noted, these include Robert Horton's descriptive finding that the number of daughter streams connected to a mother stream is between 3 and 5, and that the length of the longest daughter stream should be roughly proportional to that of the mother stream—a proportionality factor between 1.5 and 3.5.

Instead of cataloging streams in the wild, three colleagues and I used pencil, paper, and the constructal law to predict these scaling rules and to answer the fundamental question of why they emerge at all. We imagined a river basin and asked: What should it look like in order to reduce its flow resistance globally? We considered possible configurations and found that the proportion of daughter to mother streams should be 4 to 1 (not 2 to 1 or 8 to 1). Using this same approach we also predicted that their width and depth should be proportional, and that the length of the mother stream divided by the length of the daughter stream should be 2.

"Hold on," shouts the careful reader—and by that I mean you. "You derived specific values through your work—the 1-to-4 ratio of mother to daughter streams—while Horton's empirical observations unearthed a narrow range (between 3 and 5)." Good point. How do we reconcile the lack of precision and the unpredictability of finer details of a natural pattern with the deterministic constructal law that led us to our theoretical drawings? How do we account for this gap between theory and reality?

The short answer is that nature is filled with accident and variation all the time and everywhere. The same can be said of our pencil-and-paper fun; we considered 2 to 1, 4 to 1, and 8 to 1, not every daughter-to-mother ratio imaginable. More obvious is that a river basin in the Amazon forms under much different geological and meteorological conditions than one in Alaska—differences in climate, rainfall history, soil types, vegetation, and a host of other local factors affect the flow. Our knowledge of the developing internal structure of any flow system depends on two entirely different concepts: the generating principle (the constructal law) that is unique and deterministic, and the properties and external forcing of the natural flow medium that are not known predicatively and accurately at every point. The principle (that a flow system with freedom to evolve will generate designs that flow more and more easily) describes the direction of change over time in a natural world with varying conditions and constraints. If nature were a laboratory with a perfectly stable and unchanging environment, then every river basin would be identical. Instead, the river basins of nature are strikingly similar, because they have the same governing principle, which means the same rules of design and the same performance level, even though they look different. In nature, we find numbers varying between 3 and 5; given the immense diversity out there that means roughly 4. This also reminds us why flow systems continue to evolve, why there is always room for improvement.

To underscore this observation, let's return to the circulatory system. It is one of the marvels of nature that this system is so exquisitely complex that no cell is far from a life-sustaining capillary. It transports blood from the heart to this vast volume by reconfiguring its design through branching. The same with the lung: The trachea begets two bronchi, each of which branches off into smaller tubes, which branch off into two smaller tubes, etc.

Thus we see the same design we witnessed in rivers—the creation of streams and channels to improve access for flow—but with much greater precision. Instead of a range of between 3 and 5, we find the number 2 every time (until the very smallest scales). We see then that the inanimate system of the river basin and the animate system of blood vessels and air passages evolve toward the same design structure. We should add that the structure of the circulatory system might indeed be imprinted in our DNA so that the entire structure emerges in toto. But the DNA chemistry alone cannot account for the fact that the same phenomenon governs the evolution of river basins, lightning, and city traffic. The answer is the constructal law.

In addition, just as Horton found that the length of the longest daughter stream is proportional to the length of the mother stream, the Swiss physiologist Walter Rudolf Hess demonstrated in 1913 the proportionality between the diameters of the mother and daughter blood vessels, which is a factor of 2 raised to the 1/3 power. This was extended in 1926 by the American physiologist Cecil D. Murray, after whom this design rule is known as Murray's law. The ratio Hess and Murray discovered is in fact the one that reduced the flow resistance of the Y-shaped fork of the vessels and is also what we find in the real world.

Using the constructal law, our group predicted that the pairing of blood vessels (dichotomy) should occur in order to reduce imperfection when the flow is laminar. Similarly, it allowed us to predict that rivers should use the 4-to-1 ratio (quadrupling) because their flow is not smooth but turbulent. This is because the type of flow (laminar or turbulent) has an impact on the design of the branching. Most blood vessels and bronchial tubes are small enough so that the flow through them is smooth, or what we refer to as laminar flow. The branching into two (or dichotomy) should occur in this instance to reduce flow imperfection. The flow through river channels with rushing water is much bigger and faster, and consequently their flows are turbulent. Thus rivers tend to branch with a 4-to-1 ratio.

So we see in the constructal law the never-ending movie of life in action. From the man-made flows to all the other animate and inanimate flows of nature, it provides, for the first time, an understanding of how seemingly disparate phenomena are governed and united by a single principle of physics. By refocusing our attention on how things look—on their evolving designs that are the morphing boundaries of their flow systems—the constructal law reveals, predicts, and explains design in nature. It shows us how the governing laws of the universe, such as the laws of thermodynamics, work with the universal tendency to flow with configuration in order to create the pulsating, evolving designs we see all around us. It allows us to see the predictable pattern in what we had long considered just cosmic coincidence.

I have not only discovered this fact, I've lived it. When meat began to disappear from shelves in Romania during the 1960s, my father, a veterinarian, had a solution. He hatched chickens.

He had a light box that illuminated the inside of the egg so we could make sure the embryo was developing. As a teenager, I stared in awe and wonder at the growth that unrolled before my eyes each day, as the vasculature grew and spread tightly on the inside surface of the shell. I also noticed that the design I was seeing was the same as that of the river basins on the colored maps I was drawing in school. Where the chicken embryo was evolving on the inside of the sphere, the Danube basin had evolved on the outside of the spherical Earth.

Back then, I considered these similarities cool correspondences, nice ideas. Now I recognize that my father's light box was illuminating the design all around us. I am also able to see that the Earth with its river basins and other "basins"—of atmospheric, ocean, and air traffic circulation—is a vasculature woven on top of and through another spherical surface of life. So life is flow, life is movement, life is design.

CHAPTER 3

Animals on the Move

If we could rewind the tape of evolution and start it all over again, would things look pretty much the same or radically different? Would fish and birds still look like . . . fish and birds? Would dogs and cats still walk the Earth? Would human beings still rule the world? Or would our planet be populated by a mind-blowing menagerie of exotic creatures that would make the most unbridled works of science fiction seem tame?

Put another way: Should biological life have evolved as we know it or is our world just one possibility, the way the dice just happened to fall?

The late Harvard paleontologist and evolutionary biologist Stephen Jay Gould famously argued that starting over would produce far different results. In his seminal 1989 book, *Wonderful Life: The Burgess Shale and the Nature of History*, he wrote:

> I call this experiment "replaying life's tape." You press the rewind button and, making sure you thoroughly erase everything that actually happened, go back to any time and place in the past—say, to the seas of the Burgess Shale. Then let the tape run again and see if the repetition looks at all like the original. If each replay strongly resembles life's

> actual pathway, then we must conclude that what really happened pretty much had to occur. But suppose that the experimental versions all yield sensible results strikingly different from the actual history of life? What could we then say about the predictability of self-conscious intelligence? or of mammals? or of life on land? or simply of multicellular persistence for 600 million difficult years?

Gould's point was that a high level of chance has been involved in determining which organisms have survived and evolved over the course of Earth's history. This is likely so. From the meteors that caused mass extinctions to the vagaries of local conditions that rewarded particular adaptations in certain species, the unexpected and unpredictable have left indelible marks.

Nevertheless, increasing numbers of biologists are acknowledging that there are some boundaries—structural constraints and organizational possibilities—that reduce the pool of potential outcomes. There is a growing awareness that some general design rules should always govern the form of any animal life. An assumption behind natural selection, after all, is that some designs work better than others. What it doesn't tell us is: What are the principles that make them work better? What does "work" mean? What does "better" mean?

Those questions have long hung in the air, creating vast pools of observations awaiting explanation. As we noted in the preceding chapter, scientists have known for quite some time about the predictable branching pattern of blood vessels and river basins. What they couldn't figure out was why this should be so. As a result, they have avoided the question or, like Gould, focused on the influence of chance and nondeterminism. Taking a step back, we can see that this is no explanation at all. Chance and accident are the opposite of rationality; they are not knowledge but an acknowledgment of its absence. "Chance" is a code word for saying there is too much conflicting data, too many variables for us to make sense of the whole. It is an admission that we cannot see the pattern, which is the opposite of randomness and noise.

Human beings, however, abhor uncertainty, so we have transformed this intellectual impotence into the certainty of doctrine and dogma. Our ancestors ascribed much of what they couldn't explain to the actions of invisible forces, divine and otherwise. Presto, we had the answer for everything we couldn't explain. Even the contradictions—Why would a benign God allow suffering?—could be dismissed with the knowing claim that He acts in mysterious ways. Even as modern science has developed the tools to account for large swaths of natural phenomena, it has embraced the notion of nondeterminism to provide a sense of order, and control, over still-puzzling forces. It is an explanation that explains very little, turning mystery into science. Old habits die hard.

It was the best we could do—until now. Just as Newton decoded the once obscure laws of motion and modern medicine has shown that antibiotics are a far more effective treatment for infection than bleeding, the constructal law is another step in our ongoing effort to understand the world around us. It reveals that design in nature is not the result of chance but of a universal law.

In this chapter we will see how the constructal law provides a very different answer to Gould's fundamental question: If we rewound and replayed "life's tape," the evolving designs of animals (and everything else) would not be radically different. We will do this by examining the three main types of animal locomotion: swimming, running, and flying. This was one of the first areas I explored after discovering the constructal law because it directly addresses its fundamental tenets. Locomotion is movement. If the constructal law is truly a principle of physics, if flow access is the key to design, then I should be able to use it to predict not only the designs of inanimate phenomena—such as river basins, lightning bolts, and lava flows—but also the design of animate phenomena such as fish, land animals, and birds. In the process, I would have strong evidence for a unifying theory of design in nature.

Animal locomotion was also a fruitful area of study because the prevailing view in science is that unbridgeable differences exist among these three forms of movement. No one would confuse a shark slicing through the water with a rabbit hopping across the ground or a hawk gliding through the air. The constructal law allowed me to challenge this position by casting old questions in a new light. Instead of focusing on all that separates these different travelers, I zeroed in on the defining characteristic they have in common: All are *vehicles* for moving mass (their bodies, and what flows through their bodies). Thus, all should have evolved in strikingly similar ways to facilitate their flow of mass across the landscape.

Life is movement. Every living system performs better when the power that is required for maintaining its movement is minimized. Just as lightning bolts and river basins should generate treelike structures to reduce thermodynamic imperfection and increase flow access, animals should have evolved to cover a greater distance for less effort, which means per unit of useful energy derived from food.

This should be true in every respect, allowing us to predict everything, from the size of their hearts and the shapes of the blood vessels to the frequency with which they move their tails, legs, or wings to the paths they cut across water, ground, and air. The traits that have emerged, the evolutionary changes that have persisted as well as the behaviors that are learned, should facilitate flow. And, if better flow is the fundamental tendency that accounts for shape and structure, then we should see the limits of accident and chance and the power of predictable pattern.

Here's what I found.

Animals move in such seemingly different ways that scientists have long considered the three main types of locomotion to be distinct. Runners and fliers have weight, for example, whereas swimmers are neutrally buoyant. The wings of birds are structurally different from the limbs of antelopes and the tails of fish. The flapping motion of a bird's wings is unlike the hopping motions of the legs of a land animal and the undulating body of a swimming one. Birds and fish in cruising mode seem to move at a constant altitude or depth, whereas runners hop up and down. Hitting the ground during running is far different from moving against water.

Complicating the picture even further is the great diversity of body sizes, shapes, and speeds found in even a single form of locomotion. We see large and small birds and fast and slow ones; birds that walk a lot and ones that do not; birds that fly alone and those that fly in flocks. Comparing the buzzing wings of a mosquito to the majestic flight of a great blue heron would lead most people to conclude that very different processes are at work. And, let's be frank, scientists have furthered this line of thinking because diversity means that the expert on butterflies has no reason to fear the fame of the expert on blackbirds, much less the specialist on fish. The distinguished professor of aeronautical engineering does not have to give credit to those who developed his science under the name of shipbuilding centuries earlier. Diversity is also very lucrative. Jobs, salaries, prestige, and opportunities in science—with its area-specific language, concepts, books, journals, libraries, university departments, academies, and awards—depend on specialization.

Examined in toto, the design of nature contradicts this approach. Numerous investigators have found that there are strong convergences in certain functional characteristics of swimmers, runners, and fliers. Just as scaling laws inform the structure of river basins, blood vessels, lungs, and a host of other phenomena, predictable patterns appear across the board in animals. Many of these involve the strong correlation between an animal's size (its body mass) and its movement. Broadly speaking, the correlation is this: *Larger animals are faster, their bodies undulate less frequently, and they are stronger (that is, their muscles exert larger forces) than smaller ones.*

Consider once more the mosquito. It may flap its wings up to 1,000 times a second just to move a few meters. The great blue heron, by contrast, flaps its wings leisurely every few seconds at cruising speeds of between 20 and 30 miles per hour. Similarly, the guppies we keep in our fish tanks must move their tails rapidly to travel across the tiny distances of their watery domains, while blue sharks can reach speeds of up to 25 miles an hour by undulating their tails with long, powerful sweeps. Size really does matter.

What is most astonishing is that this correlation between body mass and movement holds true not just for every group of similar animals—all swimmers, runners, or fliers—but uniformly across the animal kingdom. Thus, the stride frequency of land animals scales with approximately the same relation to body mass as the swimming frequency of fish; the speed of running animals scales with approximately the same relation to mass as the speed of flying birds. Put another way, if we know an animal's body mass, we can calculate how frequently it swings its tail, moves its legs, or flaps its wings. In addition, the force output of the muscles of swimmers, runners, and fliers can be calculated from their weight: It is, for all of them, roughly equal to twice their body weight.

In an attempt to explain these consistent features of animal design, biologists have concentrated on potentially common constraining factors, such as muscle contraction speed, or structural-failure limits. These findings are descriptive, not predictive: They tell us what we see but not why this should be so before we see it. The constructal law provides meaning to this experience, enabling us to discover the relationship between mass and movement, using theory—a purely mental viewing—to predict what their designs should look like.

We start with a basic fact: It takes fuel or food to produce the work that powers every engine. The fuel or food generates an amount of heat. A significant fraction of this heat—called *useful energy*, available energy, or exergy—is *in principle* available to be converted into work. The bad news is that neither animals nor heat engines can fully convert the useful energy into work.

Some of the useful energy is lost because of many features of imperfection (flows that overcome resistances, heat currents that flow across finite temperature differences, etc.). This happens everywhere, before the animal or engine produces work from useful energy (exergy), and after (we will discuss this in greater detail in chapter 10). With the produced work, cars, people, and birds battle wind and gravity among other things; the water in rivers rubs against the hard earth and against other obstacles that would slow it down. This is how the precious fraction of useful energy that had become work is ultimately destroyed. All of it.

A well-designed flow configuration cannot obliterate imperfection. But it can reduce its global effect so that more useful energy is made available for moving the mass on the landscape. This is achieved by a better and better distribution of imperfections. To evolve toward a balance of the various imperfections, the components of the flow design must be distributed in certain ways. A river basin, for example, configures and reconfigures itself so that the water is discharged more and more easily from the entire plain to the mouth of the river. The bifurcated structure of lungs, the round tube shape of pipes, and the cracking pattern of drying mudflats are all designs that distribute their resistances so that globally the flow system becomes less and less imperfect.

Animals travel on the surface of the Earth as do rivers, winds, and ocean currents (Figure 19). All are engines that generate work to move mass in an environment filled with things (brakes) that oppose their movement. They move in different ways for different purposes, but the effective expenditure of useful energy is important over a lifetime. Like everything that flows, animal locomotion represents the tendency of moving objects to overcome obstacles, chiefly from gravity and the friction against air, water, and land.

The constructal law predicts that, if we rewound Gould's tape of life, fish, terrestrial animals, and birds should always manifest designs that allow them to move their mass farther for a given amount of useful energy that is derived from food. More power, more speed, farther, faster; these are measurable manifestations that invoke the word "better." This is the time arrow, the design direction of all the other flows that cover the Earth.

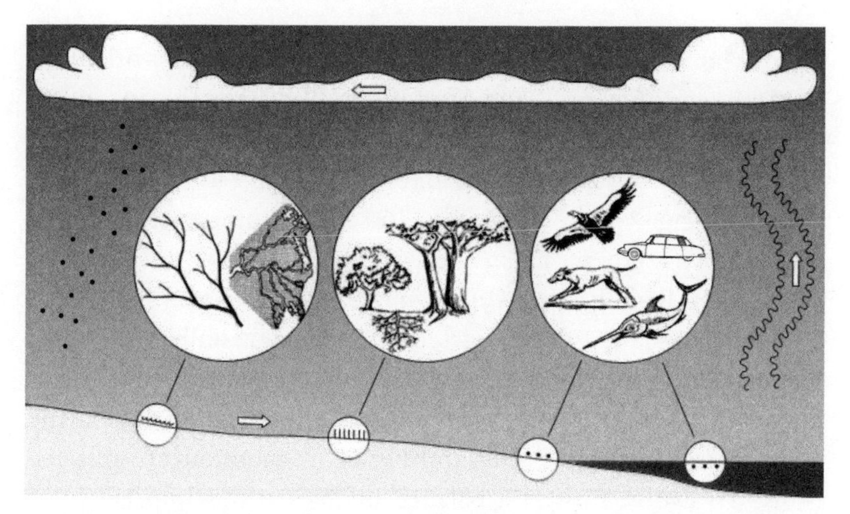

Figure 19. Several configurations that facilitate the circuit executed by water in nature: raindrops, tree-shaped river basins and deltas, trees and forests, the flow of animal mass (swimming, running, flying), and turbulent structure.

As is customary in science, for our analysis of locomotion we group animals together by basic body types. In the simplest model possible (the one that represents all the bodies), the animal body has a single length scale (L_b) and a body mass scale (M). First, here is what "scale" means. To measure the length of a housefly, we use the unit millimeter. To measure its weight, we use milligrams. This same scale applies to other insects. The L_b for a sparrow is measured in centimeters and its M in grams. The same is true for hummingbirds, so that hummingbirds and sparrows are said to have the same scale. Moving up in size, the L_b of a goose is measured in meters, its M in kilograms. Inside each scale we find a large number of animals and things that line up to be measured in the same way. For example, the kilogram scale of mass unites the goose with the duck, owl, vulture, and toy airplane.

This concept of scale is important because it underscores the fact that we are examining broad phenomena. Some sparrows fly faster than others; an obese man cannot run as fast as an Olympic sprinter—a myriad of factors, including cold and warm habitat or a weakness for ice cream, determine variation within a group. Nevertheless, the entire community of sparrows exhibits predictable flight characteristics just as, broadly speaking, human beings move their legs at a predictable rate and run at a predictable speed.

Now let's focus on an imaginary flying body, using the constructal law to predict what its design should look like to reduce the effects of thermodynamic imperfection—as well as the broader prediction that animals are built for movement. Just as river basins put the right channels in the right places to move more water for less useful energy, flying bodies should flap their wings with the right rhythm to enable them to achieve the right speed to move their mass a greater and greater distance.

We note that a flying body is a study in deception. At cruising altitude it appears to glide across the sky, riding gently on the wind, straight as an arrow. This lovely image is, alas, an illusion. Its passage is not a steady movement at a constant altitude. Its trajectory, instead, is a saw-toothed horizontal line with a tooth size dictated by the flapping stroke (Figure 20). It rises with the downward stroke that works against the force of gravity and then

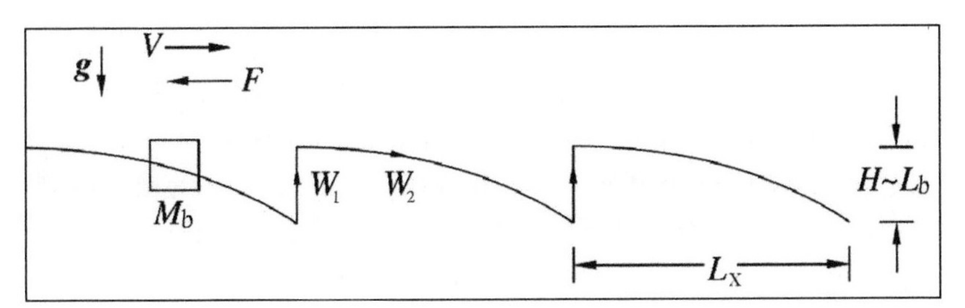

Figure 20. The periodic trajectory of a flying animal shows the factors considered in predicting animal locomotion from the constructal law. The saw-tooth pattern results because flying velocity (V) is composed of alternating work done to overcome vertical loss (W_1) and to overcome horizontal loss (W_2). W_1 is found by multiplying body mass (M), gravity (g), and the height the body falls during the cycle (H), the latter of which scales with body length (L_b). W_2 is the product of the force of air drag (F_D) and the distance traveled per cycle (L_x).

falls as the effect of this effort wanes. The essential point is that as its cruising speed increases, this vertical loss decreases but its horizontal loss increases due to mounting air friction. To see this for yourself, next time you're a passenger in a car going 30 miles an hour, stick your hand, tilted slightly up, out the window. In addition to the resistance of the wind, you feel that same wind lifting your hand (which is acting as a makeshift wing). Now have the driver increase the speed to 60. Even as you feel the wind pushing your hand back, requiring more force to keep it steady, you also feel much greater lift.

To fly at a constant altitude, a body spends useful energy to overcome vertical and horizontal loss. Neither loss can be avoided completely. However, the constructal law predicts that they should be balanced against each other so that their sum is made smaller and smaller through the selection of a rhythm in which the work of repositioning the body vertically is matched by the work of advancing the body horizontally. Balance should be achieved by wing flapping such that the flying speed is just right. This special distribution of imperfections is flight itself.

Flying, then, is a rhythm, a sequence of beats tapped out by the wings. For example, if you think of a bird, during a cycle—defined as one downward and upward wingbeat—the bird must perform work in two ways: in the vertical direction (W_1) and in the horizontal direction (W_2). At cruising altitude, the vertical work necessary in order to lift the body back to a height equivalent to its body or wing length scale (L_b) is $W_1 \sim MgL_b$, where Mg is the weight of the body, g is the gravitational acceleration (9.81 meters per second squared), and the symbol ~ means "of the same scale as" or "approximately equal to."

Meanwhile, horizontal work is necessary in order to penetrate the surrounding medium (air). This work is equal to the drag force (F) times the distance traveled during one cycle of wing flapping (L_x), namely $W_2 \sim FL_x$. The horizontal travel L_x is equal to the cruising speed V times the timescale of one cycle.

Combining these formulas, we find that the total work spent

to cover a distance is the sum of two losses: the vertical work done per unit of travel and the horizontal work done per unit of travel.

The constructal law predicts that the design of birds should reflect the tendency to mitigate these two losses. How should it do this? Given that the mass of the flying body is fixed, it should lift itself at the right rhythm to achieve the right speed to minimize the sum of the horizontal and the vertical loss, *for an object its size.*

Returning to our drawing board, these formulas reveal the simple scaling laws that we can use to design our imaginary bird. To mitigate these two losses, speed should be proportional to body mass raised to the power of 1/6, or $V \sim M^{1/6}$. For birds, the flapping frequencies should be proportional to body mass raised to the power of minus 1/6.

With these equations, the constructal law predicts that larger birds should fly faster than smaller ones and that heavier birds must do more work to travel the same distance as lighter ones, so that they must eat more food than lighter birds. It also predicts that larger birds should flap their wings more slowly than smaller birds or insects. This should be true for all flying bodies.

With the $V \sim M^{1/6}$ formula (Figure 21), we predict that a 10-kilogram bird should have a speed on the order of 20 meters per second. For a 1-gram insect, the same formula predicts a speed on the order of 5 meters per second. These are just scales, orders of magnitude for vultures or mosquitoes; they are approximate but *correct.* The scatter of the speed data in each group (insects, birds, airplanes) can be discussed further based on differences in body shape, wing slenderness, and lifestyle (terrestrial versus migratory birds).

When I tested my findings in the real world, I found that these predictions agreed well with observations over the entire range of flying bodies—insects, birds, and airplanes (Figure 21). That is, the secret of the design of flight can be explained completely by looking at how every body balances thermodynamic imperfections to achieve flight.

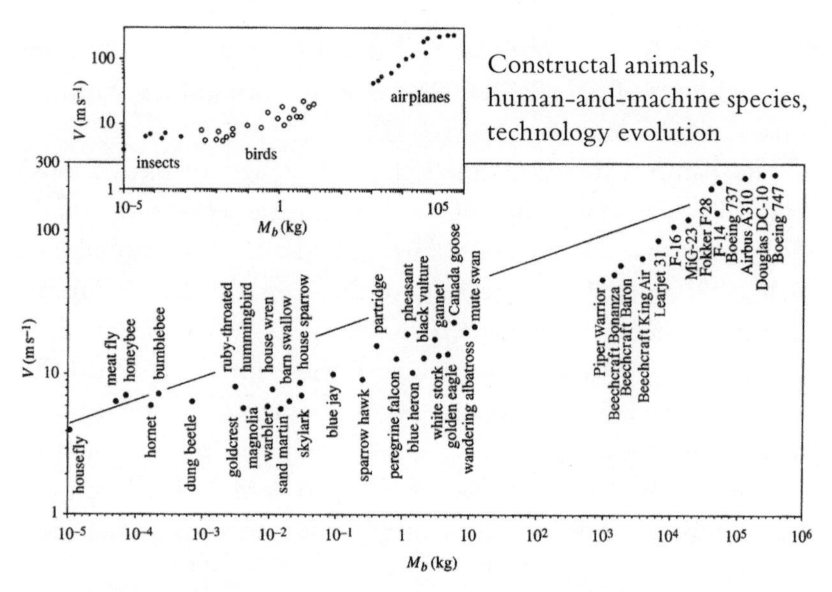

Figure 21. The characteristic speeds of all the bodies that fly (insects, birds, human-and-machine species) next to the constructal speed versus body mass raised to the 1/6 power.

I wrote the constructal theory of animal flight on a whim, while finishing the manuscript for my 2000 book, *Shape and Structure, from Engineering to Nature.* In September 2004, I was invited by Professors Ewald R. Weibel and Hans Hoppeler, from the University of Bern, to present the constructal law at a biologists' conference in Ascona, Switzerland. I used the prediction of flight as one of the many examples of how the constructal law accounts for design generation and evolution throughout nature, both animate and inanimate.

Another speaker at this conference was James H. Marden, a biologist from Pennsylvania State University. During the coffee break following my early-morning lecture about flight, he said we should try to predict the scales of running in the same way. We accomplished this, with pencil and paper, before the lunch break.

We found that if we treat running in the same way as flying—

as an effort to move efficiently against the forces of gravity and air and ground friction—we can also predict the speeds and stride frequencies of all runners. Running is, essentially, a form of jerky flight. Instead of flapping their wings to lift their bodies, runners use their legs to spring off the ground (Figure 22). At their apogee, runners' legs are airborne; they are flying ever so briefly. Like birds, their trajectory is saw-toothed, closer to a cycloid. Also like birds, runners encounter two losses of useful energy: vertically, as gravity pulls them back to Earth, and horizontally, to overcome friction against the ground and the surrounding air.

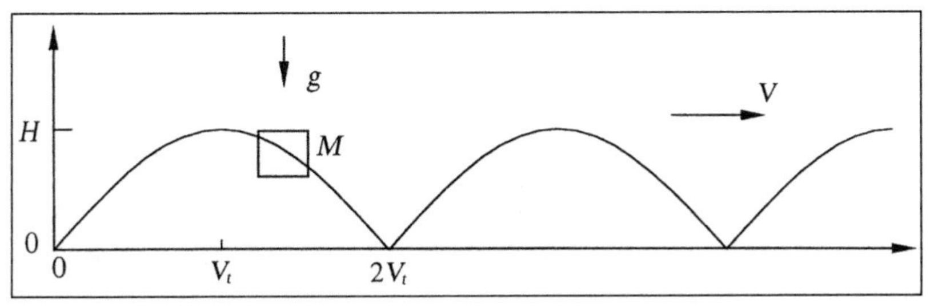

Figure 22. In the periodic trajectory of a running animal, the distance of each stride is the animal's velocity (*V*) multiplied by the time (*t*) of frictionless fall from the height of the run (*H*). Therefore, *t* is equivalent to *H* divided by gravity (*g*), raised to the 1/2 power. The stride length and *H* both scale with the body length, and the body mass (*M*) is approximated by the body density multiplied by the body length cubed.

The vertical and horizontal losses compete, and when they are in balance, their sum is lower than when they are not in balance. Here's the amazing fact that would be a strange coincidence if it weren't the outcome of the constructal law. Just as with birds, we can use the constructal law to predict the runner's speed and stride frequency if we know the body mass (*M*). What's more, this formula is essentially the same as the one we found for flight: Speed is proportional to $M^{1/6}$ and the stride frequency proportional to $M^{-1/6}$. As with birds, this scaling law allows runners to cover greater distances for a smaller amount of useful energy spent. As we will see in chapter 4, my student Jordan Charles and I applied this finding to swimmers and runners and predicted that world records must fall as champions become bigger, taller, and more slender. The stereotype of the big galoot is a myth. All other things (talent, training, etc.) being equal, bigger athletes go faster.

Another surprise comes from the calculation of the force necessary to lift the body off the ground. For both runners and fliers, the average force exerted over the stride or stroke cycle should be twice the body weight. This agrees with the force-weight measurements across all body sizes, for all animals that fly and run (Figure 23, bottom). These findings underscore the fact that what we are witnessing is not coincidence but pattern.

So far we have seen that running is similar to flying. What about swimming? Jim Marden and I thought about this for three months following our meeting in Ascona. The obvious answer is no, because the movements of the neutrally buoyant bodies of swimmers seem to have nothing to do with gravity. Before the constructal law, this view was dogma across the range of sciences and prevented the emergence of a unifying physics theory of locomotion that includes swimming.

Yes, fish are neutrally buoyant when they float in place. But when they move horizontally, they must push against something, and that something is the Earth. The ground resists everything that moves relative to it, even though swimmers and fliers do not touch it. It serves as a reference against which all moving bodies push. Archimedes declared, "Give me but one firm spot on which to stand, and I will move the Earth." He was right, and this is why swimming is no different from running and flying. Running and flying evolved from swimming as animal movement spread from water to land to the air.

This design is obvious with birds and land animals, as they fight the vertical loss caused by gravity. Birds push down and

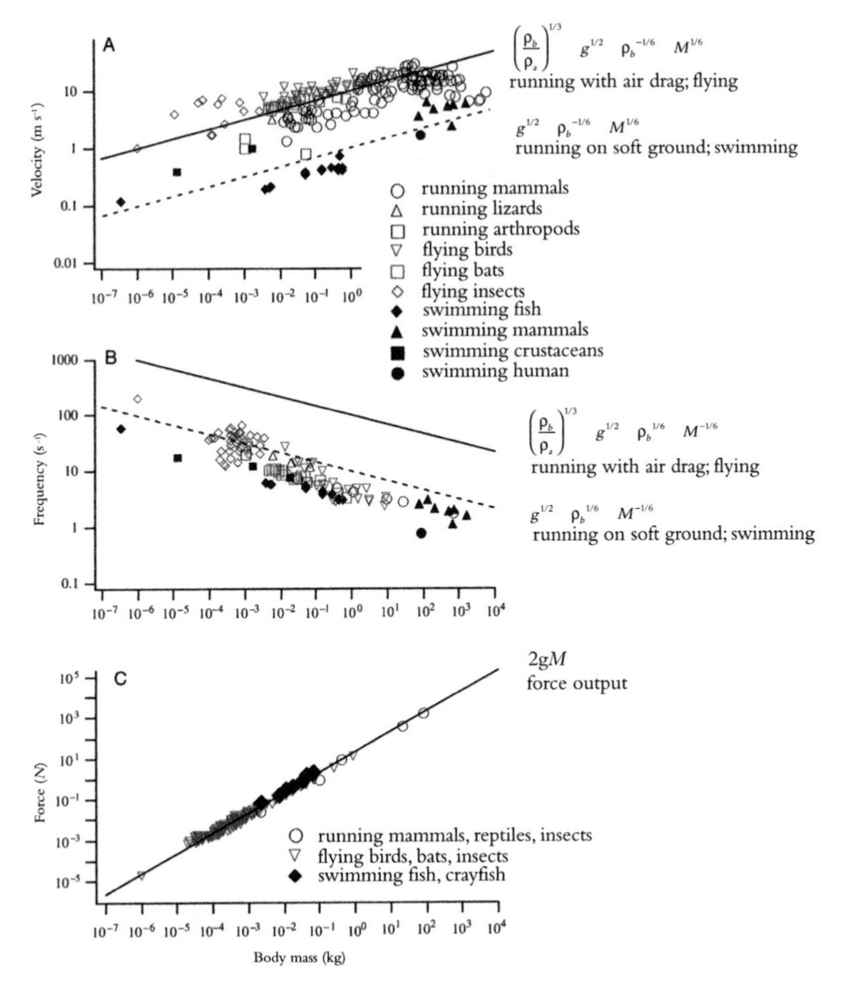

Figure 23. Theoretical predictions from the constructal law are compared with the velocities, frequencies of strokes or strides, and force outputs of a variety of animals. Solid lines in these log-scale graphs show the predicted velocity (A) or frequency (B) of animals based on body mass for flying animals or running animals where the ground is hard and thus the main frictional loss is due to air drag. Dashed lines show the predicted velocity (A) or frequency (B) of animals based on body mass for swimming animals or running animals where the ground is soft and thus the main frictional loss is due to ground deformation. A dotted line indicates the predicted force output, based on body mass (C). The theoretical predictions ignore factors between 0.1 and 10, and so are expected to be accurate within an order of magnitude.

back against the air to lift themselves, just as runners push down and back against the ground to spring forward. Swimmers—and for that matter everything that moves under or on top of the water, from boats to submarines—must also push and move their bodies relative to the ground by performing work against gravity and friction.

As high school science, common sense, and daily experience make clear, two things cannot occupy the same space simultaneously. To move forward, the swimmer must displace the water in front of it (Figure 24). To advance horizontally by one body length, the swimmer must lift an amount of water equal to its own size to a height approximately equal to its body length. This amount of water must be raised—against the force of gravity—because a net upward displacement is the only way that water can flow around an animal, or any moving object.

Why must the water move up and not down? Because the

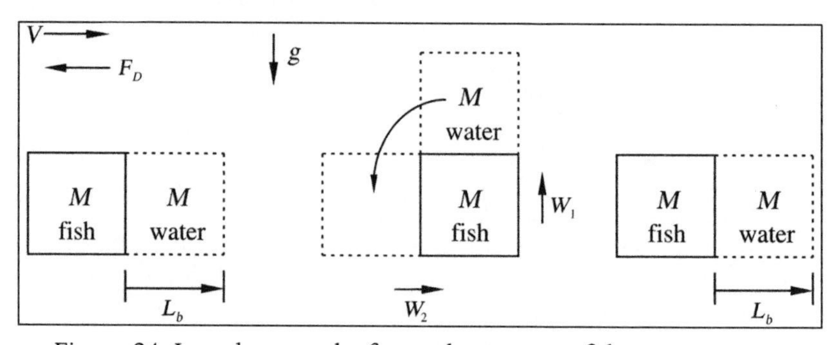

Figure 24. In order to make forward progress, a fish must move water out of the way, and the only direction the water can go is up. In order to move one body length (L, or L_b) at a certain velocity (V), a fish with body mass M_b must move an equivalent mass of water (M_w). This mass of water can then be thought of as moving downward to occupy the space now vacated by the fish. The work required to move the water mass upward (W_1) is approximated by multiplying M_b with L and gravity (g). During the same interval, the fish must do work in order to advance horizontally (W_2), that is, proportional to the force of water drag (F_D) and the distance traveled per cycle, which in this case is the body length (L).

water surface is deformable and the lake bottom is rigid. The only place the displaced water can move is to the surface, where it pushes through the air and creates a wave. This asymmetry has escaped scientists because most fish are small and swim deep. The water that fish lift is spread out over such an immense area that by the time the displaced water reaches the surface the effect is minuscule—though high-tech systems on satellites are able to detect a moving submarine by the tiny change in the surface water height over a large area.

Those of us who do not own such a detection system can observe this lifting of the surface when fish cavort near the surface and the water seems to boil. It is even easier to see as a boat slices across a lake. Its bow creates a wave that lifts above the surface of the water. You can create the same effect by filling a tub and skimming your finger across the top. The work you do to move your finger across the top causes the displaced water to rise against the pull of gravity, as it exerts two efforts: lifting water (W_1) and rubbing against it (W_2).

What has not been fully appreciated before the constructal law is that this vertical work is significant and is fundamental to the physics of swimming at all depths. Thus, even though some animals do not touch the ground, they must use it to propel themselves forward horizontally. The flapping of the bird's wings produces jets of air that eventually push against the ground and increase the pressure that the ground supports. The water lifted by the swimming fish induces a local elevation of the water surface and consequently a greater pressure on the lake bottom. The ground feels and reacts against everything that moves, regardless of the medium in which a particular body is moving. The lake bottom feels the movement of the fish.

With this, the constructal law shows how swimming is the same as running and flying. It is no surprise, then, that the predicted speeds of all swimmers are also proportional to $M^{1/6}$, just like the speeds of runners and fliers (Figure 23). The same pencil-

and-paper analysis shows that the frequency (flapping, fishtailing, stride) should vary as $M^{-1/6}$: Large animals undulate their bodies less frequently than smaller ones. Bigger fish will flap their tails less often than smaller fish; bigger birds will flap their wings less frequently than smaller ones, and bigger land animals will have less frequent strides than smaller ones.

The correlation between locomotion and body mass that we see in all animals is only part of the puzzle. The principles that govern animal locomotion also predict the flow design of other natural phenomena. The fact that gravity and lifting water are essential in swimming leads to the observation that fish advance horizontally with the same speed as the wave generated by the lifted water.

Bigger waves have higher speeds. The constructal law proclaims that we should be able to predict their speed from their size because these mindless blobs of water should also generate designs that allow them to get from here to there efficiently. Indeed, the study of water waves has shown that their horizontal speed is approximately the same as the speed of free fall from a vertical height that is comparable with the length scale of the wave. The speed at which a three-foot wave moves toward the shore is approximately the same speed at which three feet of water falls to the surface. Just as bigger animals are faster than smaller ones because they fall forward faster, taller and longer waves move horizontally more quickly than shorter ones. Furthermore, if we replace the length scale of the water wave with the length scale of the body of the fish, we obtain the speeds of all swimming bodies. This quite stunning fact is discovered when we consider that animal mass density is roughly the same as the density of water.

This similarity in density between animals and water helps us see the evolutionary connection between the animate and inanimate world. Animals are really just blobs of water; all animals came from water and spread the water on land and in the air. The innumerable waves that the winds and the ocean ceaselessly create allow us to witness evolution. They are manifestations of the process that occurred over vast periods of time to create the far more complicated structures we call fish, land animals, and birds.

Note that the predictions of constructal theory are consistent not only for waves and animals but also for man-made machines. The force–mass relation of engineered motors is the same as that of fliers, runners, and swimmers. The constructal theory of animal flight also predicts the speeds of airplanes—bigger planes are faster—and unites the animate with the inanimate. And why shouldn't they be united? They, too, confront the same problem as animals and waves: trying to move on Earth against the forces of gravity and friction. They are us—the human-and-machine species.

When we see animal locomotion as a design to move mass on Earth, other puzzles become clear. Let's start with an obvious observation that leads to a surprising conclusion: Larger animals must perform more work (force times distance) to travel the same distance as smaller ones. When the work of lifting weight (W_1) matches the work of overcoming horizontal drag (W_2), the total work per distance traveled during the cycle (L) is of the same scale as the weight of the lifted mass (Mg).

In addition, an animal's metabolic rate (the amount of food it needs to perform that work) can also be predicted from its size by using the constructal law. It, too, increases with body mass. That is, larger animals must eat more food, at a rate that is proportional to M^k—where k is the slope of the food versus M curve when plotted on a log-log graph. The constructal law showed that the exponent k must vary between 2/3 and 3/4. Thus, the constructal law allows us to predict, for the first time, a simple formula where k is not unique for determining the caloric needs of all animals.

The fact that bigger animals need more energy to move their mass than smaller ones is hardly front-page news. The headline here is that larger animals are more efficient as mass vehicles than smaller ones. Using the constructal law, we discovered this by combining our two earlier findings: The amount of food eaten per distance traveled is proportional to M^k (for clarity we set $k = 3/4$) divided by the animal speed (which is proportional to $M^{1/6}$). It follows from this that the animal food requirement per unit distance is proportional to $M^{7/12}$. Furthermore, the food required per unit distance and unit of animal mass decreases in proportion to $M^{-5/12}$ as the size of the animal increases.

For example, if an elephant weighs 1,000 kilograms (kg) and moves 1 kilometer (km), then its food intake for 1 kg of transported mass is proportional to $1{,}000^{-5/12} = 0.0562$. If the same 1 kg of animal mass is transported the same distance by 100 jackals each weighing 10 kg, then the food required by the 1 kg is proportional to $10^{-5/12} = 0.383$. What counts is the ratio between the two food requirements, namely 0.0562/0.383, which is approximately 1/7. The conclusion is that the 1 kg of animal mass travels on elephants at only 1/7 of the food cost of the 1 kg on jackals.

This fact illuminates two more big ideas. First, it provides a theoretical physics basis for the economies-of-scale phenomenon noted throughout engineering, economics, logistics, and business. The efficiency of moving something in bulk increases with size (Lorente and Bejan 2010). Second, it underscores the idea that there is a direction of evolution toward improvement of how things move. Just as raindrops occur before rivers, smaller animals appeared on Earth before larger ones—single-cell beings before elephants, mosquito-size insects before great blue herons. Using the constructal law we see the indisputable trend toward not only more movement but also more efficient movement. This time arrow is a major step that we will explore in greater depth in chapter 9.

Here is another puzzle that is elucidated by the constructal law: the sizes of organs of animals and components of vehicles. All animals have characteristic organ sizes: Larger organs on larger animals are so characteristic that all animals appear to have been constructed by assembling the same components in the same proportions of sizes. For example, the hearts of mammals weigh roughly 0.5 percent of the whole animal. Why?

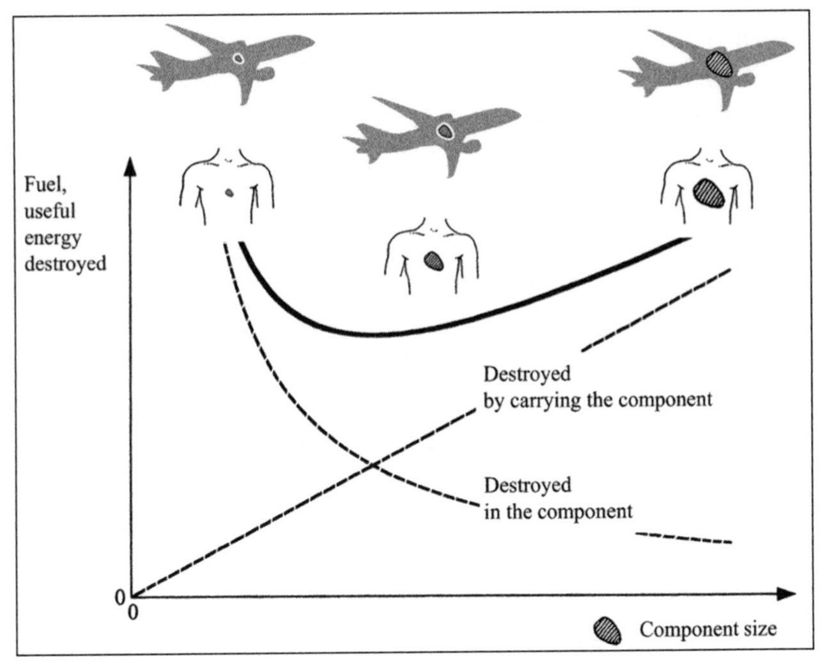

Figure 25. The organ of an animal or vehicle destroys useful energy in two ways, and both depend on the size of the organ. The cost (the useful energy destroyed) due to the flows through blood vessels and other flow constrictions decreases as the organ size increases. The cost of carrying the organ on the animal increases in proportion with the organ mass. The total cost is minimal when the organ size is such that one cost balances the other.

Imagine that the size of the heart is free to vary (Figure 25). The larger it is, the less constrictive its flow passages are, and consequently the work spent on pumping the flow is smaller. At the same time, the work spent in order to carry the organ increases in proportion with the size of the heart. The sum of the two work requirements is minimal when one work cost matches the other. This "optimal distribution of imperfection" pinpoints the organ size.

This is an important theoretical step, because it predicts the *necessity* of characteristic-size organs. When examined in isolation, the organ appears to be too small, that is, too constrictive to its flows. From this comes the frequent declarations that the organ is a mistake, that is, that "nature makes mistakes." Yes, if we were designing the heart in isolation, we might make it bigger and heavier, constructing wider pipes for the better flow of blood. But the heart is one component of a larger flow system (the animal), and we predict that it should evolve to have the right size and right weight to enhance the performance of the entire animal to move more easily on the map. These are not "mistakes." When examined as an integral part of the moving animal, the natural and imperfect organ is the one that (along with the other natural and imperfect organs) makes a good animal—an efficient construct for moving animal mass on Earth.

Because it is a principle of physics, operating everywhere all the time, the constructal law makes us think holistically. Nothing lives in isolation; every flow is part of other flow systems. If we consider a bird in flight, we can see its influence on at least three different levels at once: internally, externally, and behaviorally.

If we dissect the bird to examine its internal organs, we find that the round cross sections of its blood vessels and the shape of its heart and muscles reflect this universal tendency to facilitate the flow of blood, air, food, and stresses. Around its body, feathers minimize heat leaks and friction so that the bird can move its mass more and more efficiently. We note that many birds migrate together. They fly in a V-shaped formation because this means that only the lead bird must fly unprotected into the wind. Those behind it fly in its slipstream, where there is less air friction, enabling them to move at the same speed with less effort. This is also why the large pack of riders in a bike race (the peloton) is almost always able to reel in the few riders who break away early in the race. The individual in the group does not have to work as hard as the solo rider.

This travel pattern also provides efficiency in another way. As the lead bird flaps its wings, it pushes air down. This action creates an air wave (known as a vortex street), pushing displaced air up slightly to the outside of the flapping wing. The following bird positions itself so as to be carried along by this rising pocket of air. The same holds true for the succession of birds in the line and explains the V-like formation. The birds (and the bicycle racers) also rotate their position in the formation by taking turns at the very front. The constructal design of the formation demands this feature, in the same way that the design of the knife blade demands a sharp edge all the time. The formation is the whole animal, the "flying carpet" in which the individual fliers are the organs, and the rotation to the front is the rhythm, the intermittent breathing in the life of the whole.

The same principle explains why fish travel in schools. Each fish displaces water as it swims. Where there was water, now there is fish. As a fish swims, the water behind flows forward to fill the space the fish just occupied. The next fish, by situating itself within a body length of the fish in front of it, is carried forward by the movement of the water, allowing it to expend less useful energy to cover the same distance. (Bike racers also take advantage of this, propelled forward by the air that surges from behind into the space vacated by the rider in front.) The benefits of this effect are realized only if the bird, fish, or cyclist is not too close or too far behind—that is, within a body length. Knowing this, we can predict the design of groups of animals on the move. This mental viewing is applicable across the board. The hull of the ship and its water waves, the submarine and its Bernoulli head, the geese and the V-shaped air waves on which the flock glides, the strings of racing cyclists and cars embedded in the slipstream (another air wave)—they all lift mass and go with the flow.

In sum, the constructal law predicts complex features that have evolved as animal design. In response to Stephen Jay Gould's question, the constructal law proclaims that if the tape of evolution were rewound and if swimmers, runners, and fliers appeared again, their shapes and structures should produce the same types of speeds, stroke-stride frequencies, and force outputs of these forms of locomotion as exist today. Their circulatory systems would still have a tree-shaped design; their organs would still have characteristic sizes; and, when useful, they would follow movement and migration patterns. Because evolution has a single direction in time—to facilitate the movement of mass—the designs that accomplish this are predictable.

Determinism and randomness find a home under this same law of physics. Up close, we are awed by diversity; the differences between a duck and a goose, much less between animate and inanimate phenomena, are innumerable. But from a distance, the overall patterns of design are easy to see. Earlier we noted how observable differences in the three main types of animal locomotion have led to the prevailing view that they are fundamentally dissimilar. As we have just seen, the constructal law enables us to recognize that they are fundamentally the same. They are united by the basic tendency to balance thermodynamic imperfections, to generate configurations that balance resistances and reduce their combined effect. This is, of course, the same tendency revealed in the evolving design of river basins and lightning bolts, of you and of me.

CHAPTER 4

Witnessing Evolution

Most people think that evolution is something that we can at best imagine, because it took an enormously long time to happen. This view is wrong. We can witness evolution all we want, if we look at the changes in our technology, movement, government, and standard of living. If these evolutionary designs are hard to discern, then take a closer look at sports.

Before the constructal law, for example, the evolution of sports was unpredictable. After all, this is why people bet on basketball games and soccer matches, horse races, dog races, camel races, and all the rest. I am reminded of a joke from when I lived under communism.

> Question: Why does the model Soviet citizen read only the sports page in *Pravda*?
>
> Answer: Because that is the only section where the news was not known beforehand.

I know the joke and all its implications, unfortunately. The sports page, however, was good even then, and I read it avidly. I was raised on basketball, from preschool to the starting five in the top league and then as a member of the national select. I grew up in the company of highly gifted men who taught me that I could be like them and that it is honorable to dream of becoming better than they. In this regard, they also taught me about evolution. "Becoming better" is not just the story of athletic competition but of evolution. Everything that moves evolves in order to become better, to flow more easily across the globe.

Biologists teach us that evolution is an ongoing phenomenon, happening everywhere all the time. But they also describe it as an extremely slow process whose effects are often very difficult to witness or predict.

The constructal law allows us to understand evolution in a new light. It teaches us (1) that everything evolves, not just biological creatures, (2) that there is a predictable direction to these changes, and (3) that we can witness many entities morphing—becoming better and better—right before our eyes. This occurs every time we marvel at the tree-shaped design of lightning bolts that flash across the sky and when we watch chimneys of steam escape from pots of cooking rice.

We can also witness it in human history, including the evolution of technology and language, of science and civilization. All have morphed noticeably over relatively short periods of time—in most cases just centuries, decades, or years—to provide better access to their currents. We will expand on that insight by focusing on a subject not normally associated with evolution, sports. This is a particularly fruitful area of inquiry because it shows how the constructal law leads us to see common things in a startlingly new light. When was the last time you heard a reporter describe sports as an evolving flow system? Sports also provides a powerful example of how we can use the constructal law to predict the future, telling us why some groups of athletes are destined to triumph while others will be also-rans. Through this we will provide the first physics theory for the evolution of sports.

As we take a fresh look at our favorite games, we will also extend our discussion of animal locomotion. Athletes, after all, are also vehicles for transporting mass. They, too, reflect the evolutionary tendency to generate designs that move more easily on the landscape. And, because all locomotion is governed by a single principle, our exploration of sports will answer broader, surprisingly interconnected, questions about the evolution of biological phenomena (such as why aquatic and land animals should look so different) and technology (the evolution of the wheel).

We begin by imagining all the people who are training to become world-class sprinters. They are scattered across the globe, in middle schools, universities, and national training centers stretching from Paris to Los Angeles to Ouagadougou.

Every year, the population of sprinters selects a small sample of itself and puts it on display for the whole world to see and cheer. The small sample is selected objectively, based on physics (kinematics, to be more exact), not on wealth or political connections. The fastest are invited to run in the top races, at the world championships, and in the Olympics. Through the years, the competitors have been getting faster. Seen constructally, speed sports are a flow system that identifies, trains, and cares for a moving population of fast athletes. This flow design evolves, becomes measurably better, by producing faster competitors and record-breaking performances. But this is the trivial part of the evolution phenomenon. The subtle part is why and how the sport is getting faster.

Neither my colleagues nor I got up one morning burning to answer these questions. None of us applied for research grants to spend big money to investigate them. The questions just happened, as a result of the fact that good ideas attract interesting minds to the table. Interesting ideas are like free food for the hungry.

Since 2003, I have been offering a course at Duke each spring called "Constructal Theory and Design" with my French colleague, Sylvie Lorente. This led to our book, *Design with Constructal Theory*. Early on, Lorente suggested that students should write a research paper on a constructal design topic that was not covered by our teaching material. The term paper is now a permanent feature of the course, a source of new ideas, and a constant reminder of the exceptional level of Duke students. During spring 2008, one of them, Jordan Charles, said he wanted to examine the evolution of swimming speed records in light of the constructal law's predictions about the design of animal locomotion. Charles was not just any student; he was the starting breaststroker on Duke's swim team. His proposal reminded me of my own years as an athlete, when I wanted to learn how to become better.

I knew the answer that Charles should find. But I did not let on. Instead, I told him to compile a list of each record-breaking performance over the last century in the 100-meter freestyle swim and the 100-meter dash for men and women. I also told him to document the evolution of the *sizes* of the winning athletes.

His first set of data confirms what we already know: Speed records have been whittled down, fraction of a second by fraction of a second, over time—see Figure 26, graph A. The real news is conveyed in graph B: The new champions tend to be bigger than the previous ones. Bigger means heavier (larger M) or taller (larger L_b). The champions project themselves as a cloud of points that rises in time. The discovery is illustrated in graph C. By combining graphs A and B and eliminating the year as a variable, we plotted the champions' speeds versus their sizes. The cloud became much thinner, pointing much more convincingly to the mechanism for greater speed: *Bigger and taller means faster.* The solid lines that pass through the clouds are statistically meaningful correlations of the dots. These lines agree with the speed-mass formulas ($V \sim M^{1/6}$) predicted for all animals from the constructal law.

There is more to this discovery than the prediction that height is the advantage in body design for speed in running and swimming. We showed that the body must also be slender, with a large height/thickness ratio (S). Among athletes who weigh the

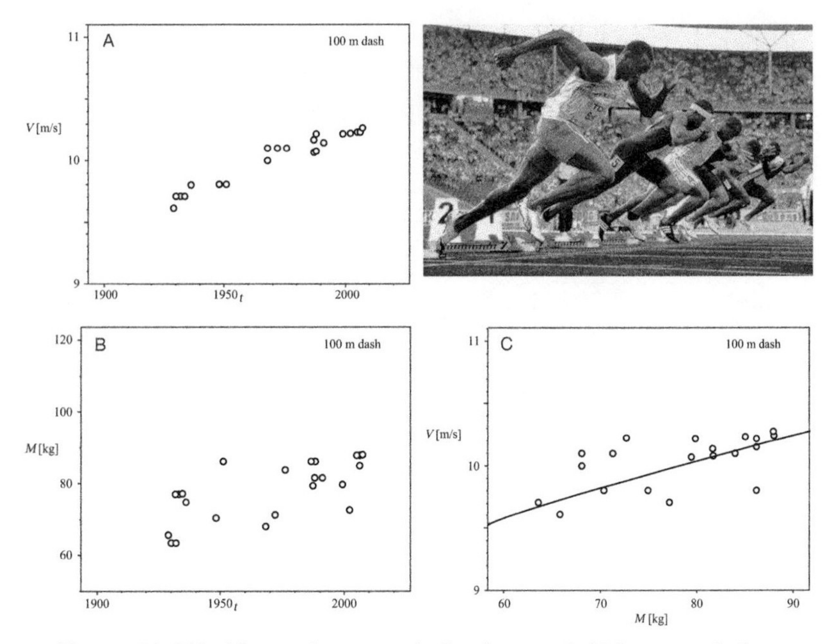

Figure 26. World running records for the men's 100-meter dash. Graph A: speed (V) versus time (t); graph B: body mass (M) versus time (t); and graph C: speed (V) versus mass (M).

same, the ones who are more slender and taller have the decided speed advantage on land and in water.

Jordan Charles and I modeled the human body as a cylinder of height H and diameter L (that is, we viewed the body mass, M, as ρHL^2), and calculated the slenderness ratio $S = H/L$ for all the record holders in the 100-meter dash and 100-meter freestyle during the last one hundred years. We showed that, as predicted, the slenderness of all these record holders increases over time, in addition to the increasing body sizes.

Now consider Figure 27, which charts the evolution of swimming records. The data shown in both figures is essentially the same. If I were a biologist, I would describe this discovery as follows. Here we have two animal populations, the world of male sprinters (Figure 26) and the world of male swimmers (Figure 27), yet their evolutionary design is the same. The biologist

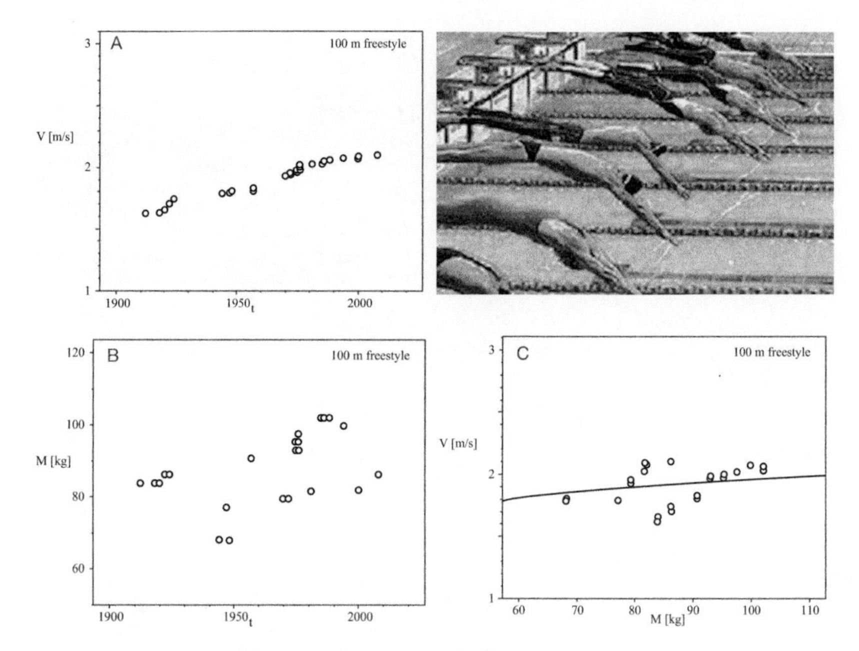

Figure 27. World swimming records for the men's 100-meter freestyle. Graph A: speed (V) versus time (t); graph B: body mass (M) versus time (t); graph C: speed (V) versus mass (M).

would also discover that we have found the same evolutionary design in female sprinters and female swimmers. The evolution phenomenon is the same, and so is the principle anticipated by the constructal law. Speeds go up in proportion with body mass raised to the power 1/6, or with height raised to the power 1/2.

Size, of course, is not everything. There is the culture, access to sports education, food, training methods and facilities, medical care, and the athlete's ambition. There is also the proliferation of performance-enhancing drugs that some athletes use. The discovery is that, all other such things being equal, size plays the same decisive role in sports as in the speeds of all animals.

Charles and I submitted our paper for publication two months before the 2008 Beijing Olympics. In the cover letter to the journal, I wrote that our paper should be published immediately because it predicts what will happen in Beijing, in both running and swimming races. The publication cycle took longer, unfortunately: Our paper appeared almost one year after the games. When it did, it was understood by the press as a successful theory to "explain" the victories of the 6-foot-5-inch Jamaican sprinter Usain Bolt and the 6-foot-4-inch American swimmer Michael Phelps. No, we did not explain; we predicted.

It created a stir because the general public understood right away that the discovery is not only about the fastest sprinters and swimmers but also about all the sports in which running and swimming are needed for winning. Almost all sports depend on speed, and therefore, to evolve, to get better, coaches must identify and nurture bigger and taller competitors. We all see this in basketball, American football, soccer, water polo, volleyball, team handball, and horse racing. Writing for the *Wall Street Journal* on September 9, 2009, Matthew Futterman called our discovery "the closest thing to a grand unified theory for the evolution of sports."

Because the law of evolution is known, it is mentally possible to flash forward the evolutionary design and to predict its future. Charles and I concluded our 2009 paper with this prediction:

> In the future, the fastest athletes can be expected to be heavier and taller. If the winners' podium is to include athletes of all sizes, then speed competitions might have to be divided into weight categories. This is not at all unrealistic in view of the body force scaling [the relation between body force and mass (weight)], which was recognized from the beginning in the structuring of modern athletics. Larger athletes lift, push, and punch harder than smaller athletes, and this led to the establishment of weight classes for weight lifting, wrestling, and boxing. Larger athletes also run and swim faster.

Better still, that paper led to another line of inquiry that addressed an obvious though taboo question of modern sport: Why is it that, from all the athletes who are fast because they are big, the fastest sprinters on the track are black and the fastest sprinters in swimming are white (Figure 28)?

This puzzle was proposed to me by Edward Jones of Howard University. This is about more than the record-breaking winners. It is about the fact that all the *finalists* in the 100-meter freestyle at the 2009 world swimming championships in Rome were white (Figure 27). Two weeks later at the world track-and-field championships in Berlin, all the finalists in the 100-meter dash were black (see Figure 26). This phenomenon is evolutional and begs to be predicted from the same principle.

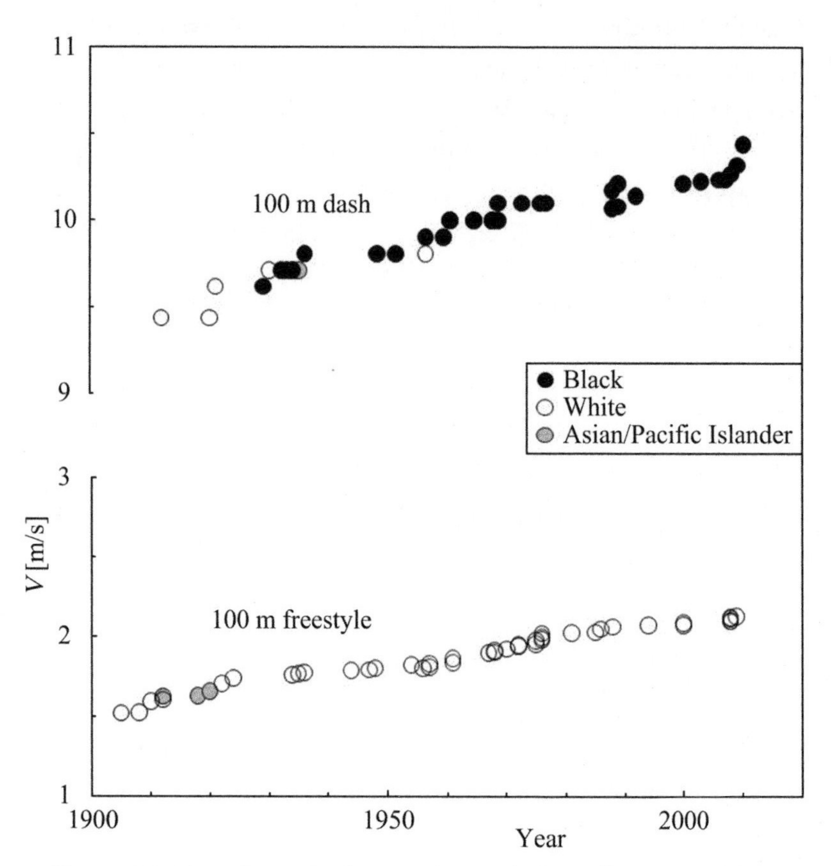

Figure 28. The effect of origin on the evolution of men's world-record speeds in running (100-meter dash) and swimming (100-meter freestyle) in modern athletics.

Jones's field of research is nutrition and how obesity is influenced by the architecture of the human body. He brought to my attention the large volume of research that documents the measured differences between the body architectures of individuals originating from various parts of the globe.

We set out to solve the puzzle of Figure 28, which, by the way, is not about gender. Female athletes dominate the speed sports according to the same pattern: black sprinters, white swimmers (Figure 29). Jones is African American, and his experience growing up as an athlete guided our inquiry very effectively. We did not get lost in the forest of political correctness. Social class and access to running tracks (and not to swimming pools) had something to do with this puzzle decades ago, but not now. Think

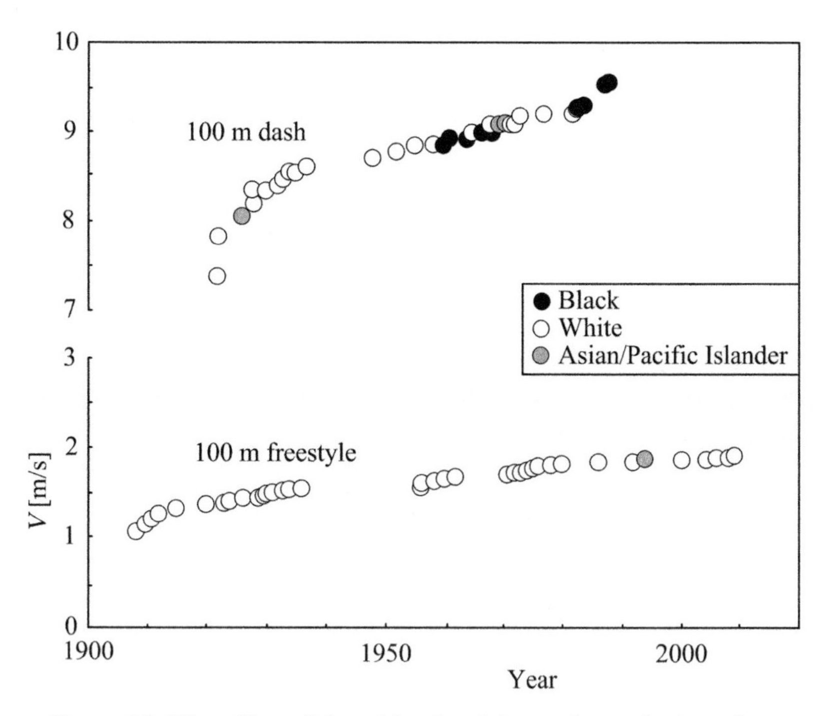

Figure 29. The effect of the athlete's origin on the evolution of women's world-record speeds in running (100-meter dash) and swimming (100-meter freestyle).

about it: Are whites disadvantaged in sprinting because they did not grow up with access to running surfaces?

The explanation derives from body architecture. The first fact that jumped out was that blacks, as a group, have body densities that are roughly 1 percent greater than the body densities of whites. This may be relevant to a comparison of the efforts needed by swimmers to stay afloat while treading water, but it cannot explain the differences in horizontal speeds in swimming. Furthermore, the data in Figures 28 and 29 are about swimming *and* running, not about swimming alone and not about swimming in place.

What we needed was a single idea that illuminated both sides of the puzzle. We scratched our heads for several weeks. But when the answer came it was obvious, beautiful, and irrefutable. Refining my previous work on animal locomotion, we realized that as animals mitigate horizontal and vertical loss as they move across water, land, and air, they are engaged in a cyclical process of *falling forward* (just like waves of water). Body mass falls down and forward, then rises again—runners spring off the ground and then fall forward; fliers flap their wings to rise in the air and then fall ahead; swimmers raise water above their bodies and then ride the wave they have created to move ahead. Bodies that fall from a higher altitude fall faster when they reach ground level. The falling speed is the same as the forward speed, and it is proportional to the body height raised to the power 1/2. Taller means faster.

The subtle point is that the body height of an athlete is not the distance from the top of the head to the ground ($L_1 + L_2$ in Figure 30). In running, the true distance is the altitude of the center of mass (L_1). Among runners of the same height, the ones with a higher center of gravity—a greater distance from pelvis to toes (longer legs)—have the advantage. Their body weights fall from a greater height (L_1), empowering them to run faster.

In swimming, the advantageous measure is L_2. Those with a lower center of mass—a greater distance from their pelvis to the

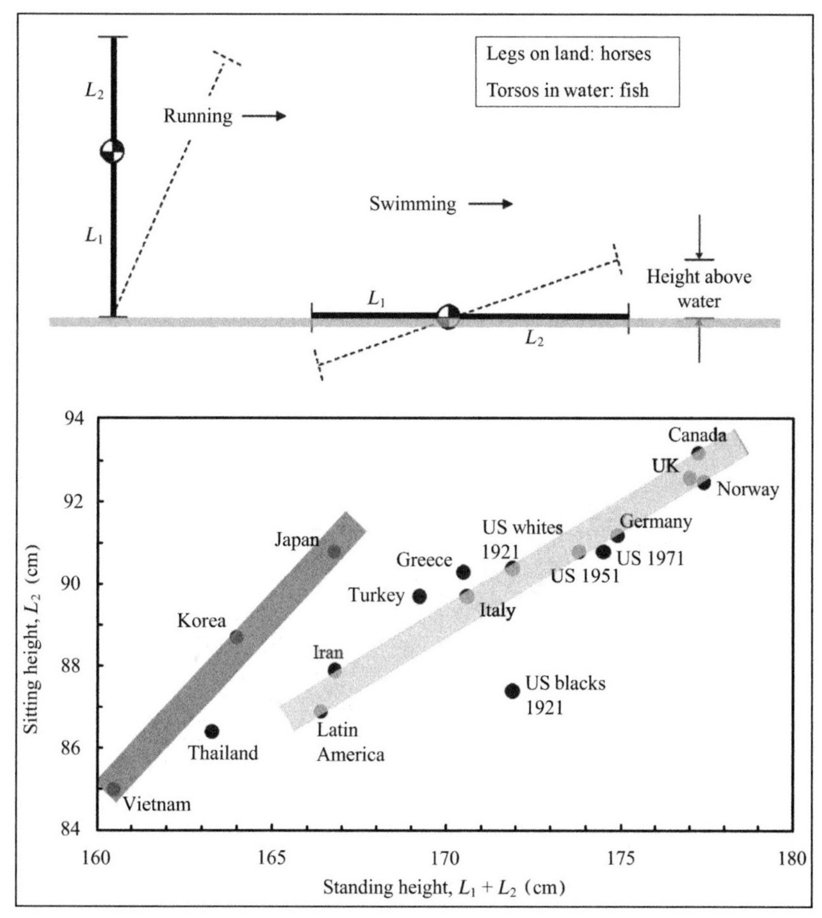

Figure 30. The correct body dimensions for falling-forward locomotion: L_1 for running, and L_2 for swimming. The overall heights ($L_1 + L_2$) and sitting heights (L_2) of seventeen groups of military men from selected populations.

top of their head (longer torsos)—have the advantage. In swimming it is the upper body (the "wave") that rises above the water line. Swimming generates a wave, so that the sport is the art of surfing on that wave. When the wave is bigger—because the torso (L_2) is longer and rises higher above the water—the wave and the swimmer go faster.

Among athletes of the same height ($L_1 + L_2$), the ones with higher centers of mass have a speed advantage in running and a corresponding speed disadvantage in swimming. The reverse is true about athletes with lower centers of mass: They have a speed advantage in swimming and a corresponding disadvantage in running.

If the constructal law prediction of sports evolution is correct, we should find that athletes of West African origin have longer limbs with smaller circumferences, so that their center of mass is higher than that of other athletes, making them faster than other runners of the same height. We should also find that athletes of Asian and European origin have longer torsos and lower centers of mass, giving them the edge in the water. This would be the single explanation of the two sides of the puzzle.

Anthropometric measurements of large populations show that systematic differences exist among West Africans, Europeans, and Asians. The published evidence is massive (we summarize a large volume of these findings in Figure 30). Measurements of seventeen groups of military men from many parts of the globe were conducted in fourteen independent studies that compared the average stature (the height of the body, $H = L_1 + L_2$) versus the average height while seated. The sitting height is not exactly the L_2 dimension defined in Figure 30, but differences between sitting heights are indicative of how L_2 varies from one group to the next.

Three conclusions follow from Figure 30. First, Asians have the largest sitting heights among individuals with the same overall height, though they tend to be shorter than other groups. According to Figure 30, Asians should be most favored among swimmers who are not tall. Second, Caucasians also line up as a monotonic relation between sitting height (roughly L_2) and total height ($L_1 + L_2$), but their L_2 is smaller than Asians'. This correlation stretches from the shorter (Iranians, Latin Americans) to the taller (Norwegians, British, and Canadians). Third, the measurements of people of West African origin fall well below those of the other groups. Their average sitting height (87.5 cm) is

3 centimeters shorter than the average sitting height of the group of men with the same average height (172 cm).

If the sitting height is an approximate measure of L_2, then the dimension that dictates the speed in running (L_1) is 3.7 percent greater in West Africans than in Caucasians. At the same time, the dimension that governs speed in swimming is 3.5 percent greater in Caucasians than in West Africans. These 3 percent differences in L_1 (or L_2) are consistent with other measurements. For example, the upper- and lower-extremity bone lengths are significantly longer in adult black females than in white females. For the lower-extremity bone lengths, the difference is between 80.3 ± 10.4 (black females) and 78.1 ± 6.2 (white females). This difference of 2.2 centimeters represents 2.7 percent of the lower-extremity length, and it is of the same order as the 3.7 percent difference between the sitting heights of whites and blacks.

In summary, 3 percent is the order of magnitude that differentiates between the positions of the centers of mass in the bodies of blacks and whites, and favors the two groups differently in the two speed sports: blacks in running and whites in swimming. For runners, the 3 percent increase in the correct height (L_1) means a 1.5 percent increase in the winning speed for the 100-meter dash. This represents a 1.5 percent decrease in the winning time, for example, a drop from 10 seconds to 9.85 seconds. This change is enormous in comparison with the incremental decreases that differentiate between world records from year to year. In fact, the 0.15 second decrease corresponds to the evolution of the speed records over 31 years, from 1960 (Armin Hary) to 1991 (Carl Lewis). The 3 percent difference in L_1 between groups represents an enormous advantage for athletes of West African origin.

For swimming, the conclusion is quantitatively the same, but in favor of athletes of European origin. The 3 percent increase in the correct length (L_2) means a 1.5 percent increase in winning speed and a 1.5 percent decrease in winning time. Because the winning times for the 100-meter freestyle are of an order of 50 seconds, this represents a decrease of 0.75 seconds in the winning time. This is a significant advantage for white swimmers, because it corresponds to evolution of the records over ten years, for example, from 1976 (James Montgomery) to 1985 (Matt Biondi).

Further support for this explanation of the speed records phenomenon is provided by Figure 29, which shows the evolution of the speed records set by women in the 100-meter dash and the 100-meter freestyle. Figure 29 for women is the same as Figure 28 for men. The female sprinters who set the records tend to be black. This trend is a bit more recent than for men, but it is just as evident. In swimming, the dominance of white women is evident throughout the modern era, just as it is for men.

This discovery is not about skin color but about body architecture. The publication of our discovery coincided with the news that Christophe Lemaitre, of France, ran the 100-meter dash in under 10 seconds. This is a first among runners of European origin. The African American sprinter Jim Hines ran this time four decades earlier. Now European commentators are pointing out that Lemaitre has unusually long legs (unusual in relation to those of other European runners). This is the advantage that is better described as having a high center of mass, one that the constructal theory of sports evolution predicted before Lemaitre's victory.

It is also an invitation to extend this line of theoretical inquiry to new puzzles that are waiting to be solved. Long-distance running is not to be confused with the 100-meter dash. We know this best from the pattern that has emerged on the winner's podium. Long-distance champions tend to be of East African origin. This was not always the case. It came as a shock in 1960 at the Rome Olympics when the marathon was won by Abebe Bikila, the Ethiopian soldier who ran without shoes. He repeated his triumph in 1964, at the Tokyo games. The pattern established itself quickly and solidly, to the point that today we expect the long-distance winners to be from Ethiopia and Kenya, and we are shocked when they are not, as when a Romanian, Constantina Tomescu, won the marathon in 2008 at the Beijing Olympics. Body architecture comes from geographical origin, and speaks of geography, of the design of movement on the globe.

The predictions based on Figures 26 through 30 are what the sports evolution theory contributes to biology. If you know sports evolution, you know that aquatic and land animals should look different. You know that the fastest runners (cheetahs, Arabian horses, greyhounds) should have high centers of mass. You know that the fastest in water should be all torso and no legs. You expect the atrophied legs and pelvis to exist inside the mammal that evolved from land to water (whale, dolphin). You do not have to kill and dissect in order to discover. You are much more powerful because you possess theory.

The constructal law's contributions are not limited to biology or the discovery that we can actually witness evolution. Because it governs everything that moves, the constructal law illuminates vast and unexpected connections. It reveals, for the first time, that the evolving designs of man-made flow systems (from technology to social systems) are also governed by this same principle of physics. That is why it should not be a surprise that my study of animal locomotion and sports, for example, has led to new insights into the evolution of one of the greatest inventions for facilitating the movement of people and goods—the wheel.

The prevailing view in science holds that the wheel is manmade, and therefore, not a natural design. This is being taught across the board. It places humans in a world distinct from and higher than that of all the other animals (and everything else) that move on Earth. Darwin must be rolling over in his grave. The constructal law challenges this perspective by revealing that the wheel is a natural design whose evolution can be predicted in the same way that we predicted the design of animal locomotion.

No invention is more closely tied to my field of mechanical engineering than the wheel. When the wheel first appeared, the movement of humanity jumped to dramatically new dimensions. It allowed us to reach higher speeds and cover longer distances with less effort. In accordance with the constructal law, it enabled us to move more mass per unit of fuel. It also marked a pivotal point in the evolution of the human-and-machine species, as people used technology to increase their movement across the globe.

Simple analysis reminds us why the wheel marked a dramatic change in how humans move on Earth—and how it tells the same story as all the other evolving designs that move more easily by reducing the effects of thermodynamic imperfection. The work (W) spent sliding a mass (M) to a horizontal distance (L) is equal to the weight (Mg) times L and a coefficient of friction (μ). With wheels placed between M and the ground, the work formula remained the same ($W = \mu MgL$) but the coefficient of friction μ decreased dramatically. The time direction of this change, from high μ to low μ, and not the other way around, is in accord with the constructal law, which states that all flow systems (including human movement on the landscape) persist in time by acquiring configurations that flow more and more easily.

Just as river basins find better and better tree-shaped flow designs every day, humans and their loads found an easier way to move mass on the map. Both reflect designs that distribute imperfection (friction, etc.) to facilitate flow. We should add that seepage in the wet mud is not eliminated by the birth of the river channel. Similarly, sliding was not eliminated by the invention of the wheel. It persists today at speeds and length scales small enough to be comparable with the movement that existed before the wheel, for example, when we slide boxes and crates on the truck bed to load and unload it. On top of the old design of movement, a better one was woven. When the old design is still a good way to flow, it persists in time. This is one reason why simple forms persist even as more complex ones evolve, why microbes and fruit flies live among dolphins and elephants, why people (myself included) still use pencils in an age of computers.

The natural emergence of the wheel design can be predicted

by using the constructal law in two ways. First, consider the evolution of the wheels made by humans (Figure 31). In the beginning, the wheel was a solid disk. The wheel and the ground made contact at the narrow strip of the rim. The stresses created by this contact were distributed nonuniformly in the disk (a). The highest stresses were concentrated in the vicinity of the contact line. Each end of the structure feels that the other end is being pulled or pushed because stresses "flow" between them.

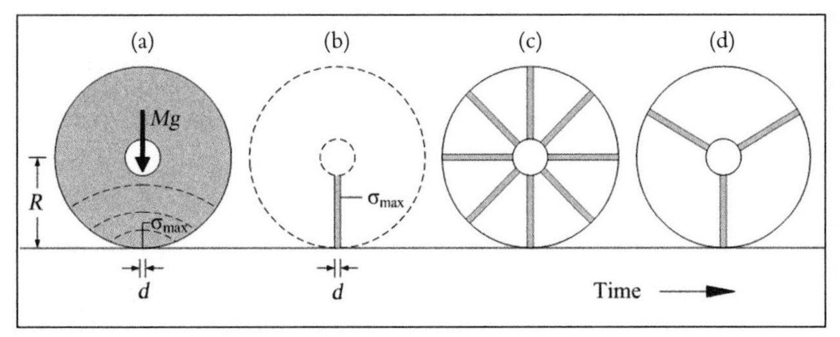

Figure 31. The constructal-law evolution of rolling locomotion, from the ancient to the modern wheel. The highest allowable stresses are distributed more uniformly, and the wheel becomes lighter and less costly in terms of useful energy destroyed in order to carry it.

Less material is needed when the maximum allowable stresses are distributed more uniformly through the support structure. The design becomes more "svelte." A single column with a uniform cross section supports the weight *Mg* while requiring considerably less material. The stresses in the column are distributed uniformly (b). The volume of the column is a tiny fraction of the volume of the solid disk.

The column is much lighter than the disk, but one column is not enough. Three or more columns, a rigid rim, and a rigid track (c) are required to prevent the body from falling. Fewer columns are lighter, and this constructal-law direction for easier movement in time is confirmed by the evolution of wheel technology in history (a) and (d).

The second way to predict the natural emergence of wheel design is to recognize the connection between the evolution of the wheel and animals. Terrestrial animals move horizontally as a rolling body—locomotion as the "falling-forward phenomenon" described earlier. Imagine the human body (Figure 32) or the

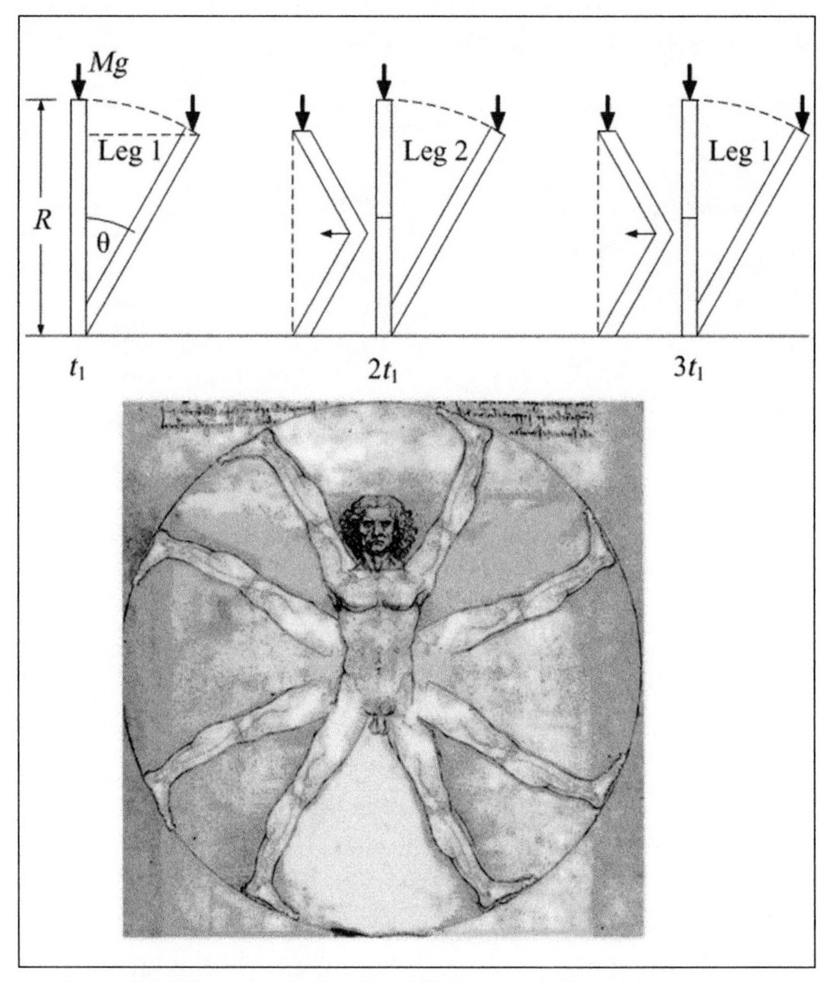

Figure 32. The animal wheel as falling-forward locomotion: t_1, the body falls forward on leg 1; $2t_1$, leg 2 lifts the body, and the body falls forward one more step; and $3t_1$, instead of requiring more legs clockwise around the body (see the human wheel), the third step is executed by leg 1, which is brought forward counterclockwise.

front or back half of a horse. If supported by a single column, the body weight (*Mg*) falls down and forward: See Figure 32, upper left, where t_1 is the first time interval of a cycle.

The order of magnitude of the speed of falling forward is the same as the speed of falling down, namely, $V \sim (Rg)^{1/2}$, where the distance above the ground (*R*) is the body length scale. Coincidentally, because $M \sim \rho R^3$, where ρ is the body density, the speed of locomotion is also recognized as $V \sim M^{1/6} g^{1/2} \rho^{-1/6}$, in agreement with the known characteristic speeds of all animals: runners, fliers, swimmers (see Figure 23), and speed athletes (Figures 26 and 27). Larger bodies move faster.

To maintain this horizontal speed, the body design requires a second column that must also have the ability to absorb shocks and to elongate itself to reposition the body weight to its traveling height (*R*). The natural design of the muscle-actuated extension of the limb is the articulation shown in the center of Figure 32 ($2t_1$). During the time interval from t_1 to $2t_1$, the second column (the "leg") lifts the body weight and moves it forward.

A third leg would continue the work of the first two, but it would increase by a factor of 3/2 the size of the organ that the animal must carry in order to move. Thus, the third beat of this rhythm is executed by the first leg, which takes the position that the third leg would have occupied (see the right side of Figure 32, $3t_1$). The repositioning of the first leg could be done clockwise (as in the human wheel, shown by this modification of Leonardo's Vitruvian man, the lower half of Figure 32), or by swinging it counterclockwise, from behind the body. The second alternative is much lighter and faster, and (in accord with the constructal law) it is the natural design of rolling locomotion.

Two columns (legs), swinging back and forth, perform the function of an entire wheel-rim-track assembly. They do it with one wheel using just two spokes and with uniformly stressed material in each spoke. No wheel is stronger and lighter than this.

The animal body is both wheel and vehicle for the animal mass that moves on the surface of the Earth. The wheel and all such "inventions" occurred naturally and are manifestations of the universal tendency captured by the constructal law.

This natural phenomenon is much more general. To see how, ask why a falling body should develop one rigid column at all. Why do bones and skeletons exist? The answer is an unexpected connection with the mechanics of deformable solid materials. When a force is applied suddenly at one point on a loosely packed material—as when a projectile strikes the ground—momentum is transmitted to the entire volume through a spontaneously generated tree-shaped network of lines of high stresses, as seen in Figure 33. Such a tree of stress lines would form inside the animal body if it were to hit the ground like a sack of packed granules. The constructal law predicts that the animal should allocate mechanical strength—more and stronger material—

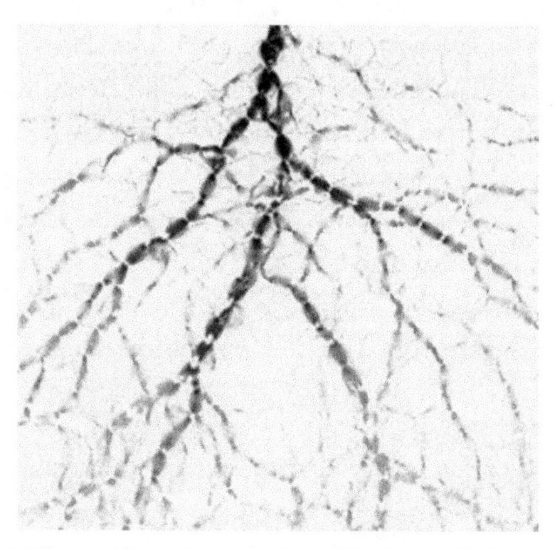

Figure 33. When a force is applied suddenly on living tissue or soil, momentum is transmitted from the point to an entire volume by a spontaneous tree-shaped alignment of grains that transmit high stresses. The living system's constructal-law tendency is to allocate stronger and more amounts of material along the channels with high stresses. These reinforcements become bones and tree roots.

along the channels with high stresses. These reinforcements, which *must* emerge, are bones and tendons. In the next chapter we will see how this same prediction applies to the roots, trunks, and branches of trees.

The animal leg is shaped like a column because it facilitates the flow of stresses between two points—foot and hip joint or paw and shoulder—not between one point and a whole volume. The leg is a column for stresses that flow between only two points because each point must be a hinge that permits free rotation around the point. When the flow of stresses is between a point and a volume, the living solid structure is rigid and tree shaped, like the rib cage, which is built as branches on two tree trunks: spine and sternum. The fact that the bone "column" is round in cross section is another matter entirely. It is a constructal requirement for a light design that is strong in bending from all possible and random directions, and it is also why all tree trunks, branches, and roots have round cross sections.

Here is another prediction—about the design of the jaw—that jumps from our minds now that we see the origin of the round cross section of the solid column that must be strong relative to random lateral bending. The dentition is the blade of a curved knife, shaped into the same U as the jaw. One could argue that the lightest such knife should be a continuous, U-shaped blade, but this blade would chip easily when stressed in ways other than tip-to-jaw compression. Much stronger (which means, conversely, lighter) is the U-shaped knife made of nearly cylindrical teeth. The front teeth are not exactly the same as the rear teeth because the loading directions in the front part of the mouth are fewer (essentially vertical compression during biting) than at the back, where the chewing represents horizontal shearing in many directions.

Nature evolved not only the design of wheel-like movement but also the design for changing speeds. Because bigger means faster, greater speed could be found by increasing the height of the body mass above the ground—the height from which the body falls forward.

Figure 34 is about the evolution of animal body movement—crawling, quadrupedal, bipedal. The snake, the horse, and the human are the best-known icons of these three types of movement. There is a time arrow to this evolution, and, if the body size is fixed, then the average speed (not just the short bursts of fast animals like cheetahs but speed over a lifetime, for moving mass on the landscape) would increase stepwise from left to right in Figure 34.

Next, assume that the body size (M) and body shape (D_c/L_c) are fixed, as in Figure 34 (b) and (c). Can this animal change speeds? The answer is yes: The animal becomes faster by orienting its longer dimension (L_c) vertically, that is, by making itself taller. The constructal-law direction is from (b) to (c), and this, too, agrees with the evolution of animal locomotion: Bipedal locomotion evolved after quadrupedal locomotion, not the other way around.

There are many examples of the animal design for changing speeds. A human has two speeds: walk and run. A horse has mainly three speeds: walk, trot, and gallop. The human and the horse increase their speeds by increasing the height from which their centers of mass fall during each locomotion cycle.

From the walk to the gallop, the horse body movement changes abruptly such that the amplitude of the jump increases stepwise. The animal body with three different designs for movement (rhythm) is like one vehicle with one engine and a gearbox with three speeds.

The evolutionary designs of nature have arrived at wheel-like locomotion and at changes in body movement that result in changing speeds. The designs developed by humans are latecomers to this long evolutionary sequence. Yet they come from the same natural tendency to move on Earth more easily. These design features are part of our own evolutionary design for moving our mass on Earth. They represent the evolution of the

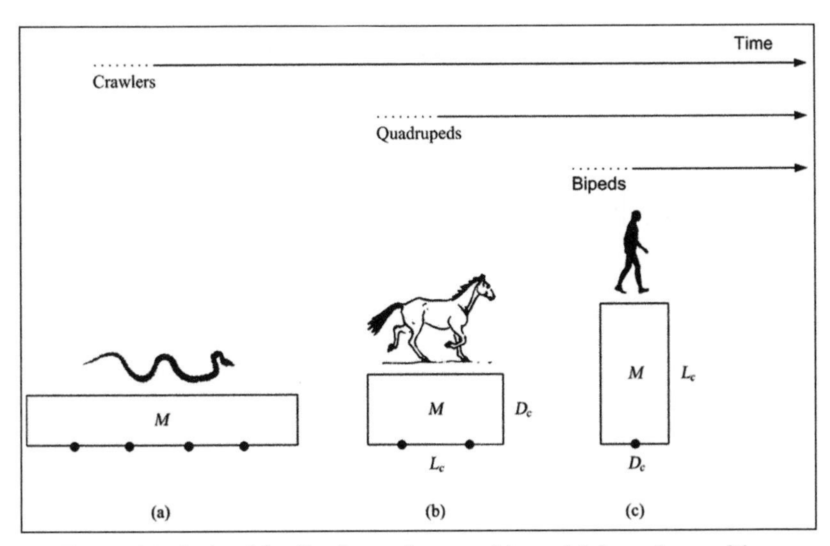

Figure 34. Animal bodies have shapes with multiple scales, unlike the single-scale (R) discussed until now. A more realistic body shape that is still very simple is the elongated body of length scale L_c and shorter transversal scale D_c (that is, $M \sim \rho L_c D_c^2$). Assume first that M is fixed and L_c is oriented horizontally, as in (a) and (b). Evolution toward higher speeds points toward locomotion designs that are taller. This means evolution toward designs with larger D_c. The time direction commanded by the constructal law points from (a) to (b), and this agrees with the evolutionary design of animal locomotion: Quadrupedal motion occurred after crawling motion, not the other way around. Crawling motion was followed by crawling plus quadrupedal, and crawling plus quadrupedal were followed by crawling plus quadrupedal plus bipedal. Each new design of motion is superimposed on (interlaced with) the existing designs. The new movement does not eliminate the old. Greater speeds emerge as the horizontal body becomes taller, that is, less slender; compare (a) to (b). Greater speeds are also attained as the body rises vertically; compare (b) to (c). In this time direction, the number of "wheels" in contact with the ground decreases. One such wheel (one black circle) accounts for two legs, as in Figure 32.

human-and-machine species, which is the same phenomenon as any other design in nature.

Engineering makes a contribution to understanding design in nature in a way that the other sciences cannot. Biologists and geophysicists argue correctly that it is difficult to witness and test "evolution" in their fields because of the enormous timescale often involved in the phenomena they study. On the other hand, by presenting animal design as a falling-forward movement akin to rolling and changing speeds, we teach an important lesson for the current debate of design and evolution in nature. We can, in fact, witness and test evolution during our lifetime by studying the evolution of our designs and technologies. These evolutionary designs illustrate the time direction of the constructal law, which unites the animate and inanimate design phenomena.

It is the same evolutionary direction and design that emerge in distinct groups of individuals that pursue the same goal: winning. The real goal is not speed, it is to win, which means to advance in society, to live better, longer, and move farther. The goal is life itself. The urge is to live. The urge is also known as the instinct of conservation (or self-preservation), which is supreme.

In the evolution of different sports groups toward a single design—and seeing how sports has selected athletes with certain characteristics—we have a parallel example that helps illustrate the evolution of different animal species into the same shape and mode of movement: the shark and the dolphin. One is a fish and the other a mammal, and the fish are much older than the mammals. Over time, however, all flow systems tend to evolve into the designs that facilitate their movement. Land mammals have different designs from fish because they move in different environments. The dolphin and whale represent a sort of evolutionary U-turn back into the water, but arrive at designs similar to fish (sharks) not by mimicking them, but because this is the constructal law's direction of body evolution for aquatic movement.

Forget biomimetics. No live thing is copying another live thing. No matter how smart, the dolphin is not copying the shark. They are different—each in the present-day frame of its movie of design evolution in big history. With the airplane, the human-and-machine species is not copying the albatross and the V-shaped flock of birds. These animals—the bird and the human-

and-machine species—could not be more different, birds versus mammals, older versus more recent. Yet the better the airplanes fly, the more they look the same and (big coincidence) they look more like the birds. They arrive at the same features because the direction of evolutionary change is the same for everything.

Through these insights, the constructal law provides us with a broader and much sharper vision of evolution. We see that it is not just a phenomenon of biology but of physics. We find that we can witness evolution by paying attention to all around us, from the evolution of the wheels on carriages to the evolution of sports.

CHAPTER 5

Seeing Beyond the Trees and the Forest

There is a beautiful stand of woods near my home in Durham, North Carolina, called the Duke Forest. Walking through it, I feel leaves crunch beneath my feet and hear birds in the air as I pass under the towering trees whose names are pure poetry: longleaf pine, yellow poplar, bald cypress, red oak, willow oak, sweet gum, hornbeam, shagbark hickory, and southern sugar maple.

It is a tranquil, calming experience. It can also be deeply instructive. As Henry David Thoreau said, "I went to the woods because I wished to live deliberately, to front only the essential facts of life, and see if I could not learn what it had to teach, and not, when I came to die, discover that I had not lived."

Thoreau's words have long resonated with me, even if his method of inquiry leaves me a little cold. I enjoy strolling through the woods, but the idea of living alone in a cabin by a pond is not my ideal. I like civilization. Creature comforts are a good thing! Nevertheless, in the years I have spent exploring the treelike patterns that occur throughout nature, I have been deeply impressed by how often poets, philosophers, mystics, and everyday people use trees—both actual and metaphorical—to express ancient inklings and hard-won wisdom about the interconnection of everything on Earth.

We illustrate our connection to one another by constructing family trees. Warriors offer olive branches to end hostilities and come together. We speak of library branches, bank branches, and branches of government to describe the tendrils of unified organizations. Every field has its branches of study—and Adam and Eve courted peril by eating from the tree of knowledge. Albert Einstein asserted that "all religions, arts, and science are branches of the same tree." I could offer many other references along these lines. The idea of the oneness of nature is as old as humanity. But until now it has remained an intuition, a hunch, a know-it-in-your-bones truth that possessed everything but scientific—that is, rational, verifiable—proof. The constructal law provides this missing link, putting meat on those poetic bones by pinpointing the principle that has long been described through metaphor. First, it reveals that for all the wondrous diversity we find in nature, everything that moves is a flow system. Next, it predicts that, given freedom, flow systems evolve over time in order to move more easily. Third, it shows how this universal tendency accounts for the patterns that we call design in nature. Finally, it illuminates the fact that all flow systems are connected to and shaped by other systems, in a global tapestry of flow.

We will continue our exploration of those discoveries by taking a look at the trees and shrubs that have long served as visible symbols of oneness. As we learn why trees look the way they do—why they must have roots, trunks, branches, and leaves of particular size and shape—we will find more evidence of how efficient, harmonious designs evolve spontaneously in the natural world. In the process we will not only feel but also understand the interconnectedness and oneness between ourselves and everything around us.

An irony of the constructal law is that it is a scientific principle that challenges scientific orthodoxy while confirming impressions of the world held by nonscientists. Like Thoreau, most people intuit the oneness of nature when they walk in the woods. Their hearts soar with a sense of the overarching "design of nature." Scientists, on the other hand, are trained to slice and dice the world. Most see the forest as a laboratory of diversity and randomness, a complicated and confusing environment they make sense of through categorization, compartmentalization, and specialization. One researcher is the master of junipers, another of the loblolly pines; all are geniuses of the genus.

On the surface, this focus on differences makes sense. The spectrum of vegetation in any forest is densely packed with distinct images: the large and the small, the slender and the robust, the soft and the hard, the large leaf and the needle, and the few and the many—all thrown together on the forest floor. Even if we look at a single species, we do not find two identical trees, two identical branches, or two identical leaves.

In challenging the scientific community's focus on diversity, it is tempting to say that it has fallen prey to that old adage about failing to see the forest for the trees. In fact, previous researchers were giving it their best shot, advancing understanding through the tools and knowledge available to them. Until the constructal law, we didn't have a scientific explanation for the deterministic design of trees and forests. Following Darwin's lead, the established view of vegetation in biology describes trees as living structures that emerge during a highly complex evolutionary process, driven by an ever-growing list of competing demands. A tree must catch sunlight, absorb carbon dioxide, and transfer water to the atmosphere, while competing for these resources with its neighbors. It must be self-healing, mechanically strong to survive strong winds, ice accumulation on branches, or damage by animals. It must survive droughts and pests. And a tree must adapt and morph, to grow toward the open space and to bulk up in places where stresses are higher.

This description is accurate, so far as it goes. What it leaves out is the principle and the physics. Why "must" the tree do all these things? Why do trees act as if they have minds of their own? And, even though no two trees are identical, why are they all treelike? It also ignores a fundamental question: Why do trees exist? The constructal law answers these questions—and offers a unified theory of design in nature—by showing us that their shape and structure can be predicted from the universal tendency to facilitate flow access.

For starters, let's acknowledge that it is hard to recognize trees as flow systems. If there's anything that doesn't seem to move, it's a tree. Despite the breeze that ruffles their leaves, trees are nature's great exemplar of, well, rootedness. Behind the bark, however, we see a different story. Trees are abuzz with activity. Just as river basins are flow systems for moving water from the ground to the river mouth, trees and forests are pumping stations operating 24/7 to move water from the ground to the air. When we start with the constructal law's first question—what is flowing?—the answer is water. The tree is a design for moving water. Beginning with the roots that pull water from the surrounding area, to the trunk that conveys water to the branches, to the leaves that release it when they open their pores to capture sunlight for photosynthesis, the design of the tree is geared toward performing this work efficiently. Indeed, when you douse your beloved camellias and gardenias with water, only a small fraction is consumed by your plants. The bulk of it is pumped back into the atmosphere.

Key to this understanding is the principle of flowing from high to low. The second law of thermodynamics proclaims that nature should manifest the tendency to move water from wet to dry both locally and globally. Trees and plants are like straws used by the drier air to suck water out of the ground. Before condemning the atmosphere as selfish user, remember that when it becomes oversaturated it replenishes the ground through rain.

The constructal law teaches us that trees and forests occur and survive in order to facilitate rapid transfer of water from the ground to the air. It improves on the Darwinian view that casts trees as individuals and separate species competing against their neighbors to survive. Taking a step back, we see human projection written all across Darwin's account. It is how people in the West tend to describe the "struggle" for life. It is hard to avoid overlaying the meaning we have drawn from our own experience onto the world. The history of science can be read as the evolving effort to replace subjective analysis with objective criteria.

Through its integrative approach, the constructal law teaches us that trees and other forms of vegetation are part of an immense global flow architecture—along with all the river basins, raindrops, and atmospheric and oceanic circulation—that facilitates the cyclical flow of water in nature. Think of it like this. In the beginning, there was water. Because of the second law of thermodynamics, water is governed by the natural tendency to equilibrate all the moisture in the environment. Because of the constructal law, a wide range of morphing and mating flow designs have emerged to facilitate that movement. That vegetation is a design for water flow is made clear by the strong geographical correlation between the presence of trees (sizes, density) and the rate of rainfall.

Trees "happen" because that is where the water is and must flow (upward), not because "trees like water." Similarly, river basins happen where the water is and must flow (downhill). Both are living systems that have emerged and evolve to facilitate the local and global flow of water. Both are manifestations of the constructal tendency to generate designs to move more mass (in this case water) on Earth. Through evolutionary history, the right designs have emerged in the right places to facilitate this flow. Cacti, for example, are relatively big and not dry, but their design is for low water transmission to the wind, an outflow that matches the water inflow associated with sparse rainfall. The water stored in cacti is flowing through the movement of desert animals that eat cacti, and in this way the cacti flow systems spread much more than they would in the absence of symbiosis with animals.

I cannot emphasize this enough, because it embodies the new perspective offered by the constructal law. In the traditional view, there are relatively few plants in the desert because there is not enough water to sustain them. This is correct as far as it goes, but it misses the crucial point that few plants are needed there because there is so little water. Why, after all, would you need a lot of water pumps in the desert? Similarly, there are usually more trees in a valley than on top of a mountain because the volume of water rises as the elevation decreases, meaning there is more water in the valley to pump back into the atmosphere.

Governed by the constructal law, rivers occur where there is water that must flow; lightning bolts happen when charged clouds must discharge electricity, and trees grow in greater abundance where there is more water in the ground than the air. Animals occur where there is water to flow, as animal mass flow. It's no coincidence that animals hover close to sources of water. Almost all animals are composed mostly of water. You can think of animal mass flow as this mass of water moving across the landscape. Rivers, lightning bolts, trees, and animals are *designs* that emerge to handle the *currents* that flow through them and along with them. They do not exist in service to themselves but in service to the global flow. Though we can look at flow systems in isolation, they work hand in glove with all that flows around them, evolving to enhance the movement of everything on Earth.

In addition, we find recurring patterns across all phenomena. Until the constructal law, no one linked the evolution of inanimate and animate systems. A river has no DNA to pass on to the fish that swim in it. Yet we predict and find that the design of river basins and circulatory systems should have tree-shaped architectures. This commonality can be understood only when we recognize that their evolutionary history has been governed by a single principle: the constructal law.

With these ideas in mind, let's take a closer look at how the constructal law predicts the design of a tree—from its roots and trunk to its branches and leaves. We begin by noting that the tree has to handle two types of flow. The first is the movement of water from the ground to the air. The second is the flow of stresses caused by the wind. Thus, a tree should have a special architecture that provides access for the water coursing through it and mechanical strength against the winds that buffet it. Given these requirements, and working at our drafting board with pencil, paper, and the constructal law, we ask: If we set out to design a system to facilitate these two types of flow—with only a blank page and no sense of trees—would our drawing eventually look like a tree?

Following the flow of water, we'll start with the root and consider how it ought to handle this flow. Our drawing of a tree root must have a porous body to allow for two types of water flow: transversal (from all sides), so that water can enter the system from various depths, and longitudinal, so the water can move up from the ground. The longitudinal flow (the through channel) has less resistivity than the transversal flow.

As we get closer to the ground, we must make our drawing of the root wider to handle the increasing quantity of water entering from the various access points below. Our root can have any imaginable shape: cylinder, cone, a needle with an infinitely sharp tip, a finger with a round tip, and many other shapes. Which one is best suited for the flow that tries to move through the root?

Of all possible shapes, the cone or carrot in Figure 35 (b) offers the least resistance as a whole. And when we enter the forest, we find that all roots are essentially carrot shaped; they are tapered. Not surprisingly, this is the same drawing we arrived at in chapter 2 to predict the design of river basins, which is also a root system in which water seeps transversally from the banks into the main longitudinal channel.

Now that the water has flowed into our cone-shaped root, what is the best way for it to move up and out? Our drawing revealed that a single conduit with fixed volume and length offers greatest access when it has a round cross section and a uniform diameter. The round cross section is also good because it offers the most resistance to bending in all possible directions. This is why we find the same design in our blood vessels, for example. Back in the forest, a little digging shows that roots have round

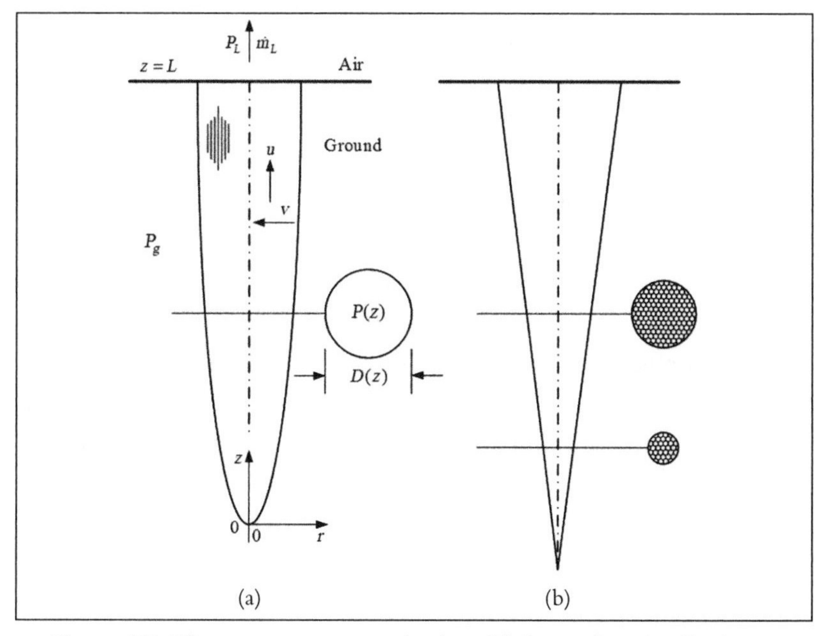

Figure 35. The root as a porous body with larger longitudinal permeability: (a) arbitrary shape of a body of revolution; (b) constructal root design. The natural design of the root resembles (b) because this is the shape that offers greater access to the ground water into the root and up to the trunk. The circles represent the shape of the cross section at the distance z from the tip. In Figure (b), inside the root cross section we see the cross sections of the bundles of tubules.

cross sections and the tubes within them are round and have a uniform diameter.

Moving from the individual roots to the entire root system, the constructal law predicts that it should have a treelike architecture because this is a good way to provide flow access from a volume (the soil) to a single point (at ground level). Wherever we find point-to-area flow that's sufficiently fast—whether it's a river basin or a lightning bolt, the air passages in our lungs or the neurons in our brain—we should also find a treelike structure because this is the efficient flow design. The threshold that defines what "sufficiently fast" means is specific to the environment of each flow system, but conceptually it is as predictable as the birth of the first fluid roll (eddy) we showed in Figure 10. Thus we find that of all the possible root designs, the one that has evolved is the one we would have designed to facilitate the flow of water. That is why treelike designs are predictable.

But what about mechanical strength? Here the constructal law expands our notion of flow, providing a new way of understanding not just the design of trees but also that of bones, bridges, and other solid objects. Engineers and biologists have long been aware that objects must be strong enough to withstand two forces that could destroy them—the weight of the object itself and outside forces that can break them. Bridges, for example, must be strong enough to support their own weight, plus that of the objects that move across them, as well as the impact of winds and other environmental factors. The bones in our legs must be able to support our weight and the added force produced by moving, jumping, and falling.

Until now, engineers have described the action of these forces in static terms. They speak of the concentration of stresses that rest in an object and then disappear when the source of the stress (the wind that blows through branches or against cars on a bridge) is removed. In a bold move, my colleague Sylvie Lorente recognized that *stresses flow* through an object. When you pull from the two ends of a rod, the force exerted flows through it from one end to the other. One end "feels" that the other end is pulled. If the rod has a uniform cross section, the stresses flow through the entire body without strangulations because this is an efficient way to use the volume of solid material to house the stresses in the smallest, lightest body. Thus, even a solid piece of steel, for example, is a flow system, uniting its design with that of rivers, birds, and everything else whose movement is apparent to the naked eye.

Bearing this in mind, we note that most recent considerations of the structure of trees have focused on their ability to resist buckling under their own weight. But the tree's weight is relatively static; it tends to increase in proportion to its size. A far more damaging factor is the notoriously random and devastating force of the wind that is constantly imperiling the tree: What sticks out too much is snapped off. The vegetation architecture that strikes us as "design" today is the result of this never-ending assault.

Just as you can feel a knock on the head down to your toes, the wind sends shivers throughout the tree, introducing stresses that flow through it. The tree's design distributes these stresses uniformly, spreading the highest stresses (and the chance to fracture) throughout, so that each part withstands the maximum allowable stress, giving each part a maximum chance at survival. Lorente's easiest flow of stresses is the principle that explains why storms snap off both thick, heavy branches and thin, light ones. Every part is at equal risk and is equally protected. This same principle predicts the design of bones—the body's bridges for the stresses that run through it. Longer bones are round, with a uniform diameter, because this is the lightest design for distributing stresses uniformly. Bones also mushroom at the ends, because they serve as anchors for the tendons and ligaments.

Returning to tree roots, we predict and find a marvelous piece of design. The round cross sections and branching structure that facilitate the flow of water also steady the tree. If the wind blew in only one direction, an I-beam type of cross section would work better—which is what we find, for example, in the sternum of a chicken that must handle only the pectoral stress caused by the flapping of its wings. But the wind blows through trees from all directions, so the round cross section is best for distributing the highest stresses among the fibers as they are subjected to bending. Like the fingers of a hand encircling a lamppost to steady the body in a strong wind, the tree roots grab hold of the earth.

This finding sheds light on one of the most famous treelike structures in the world, the Eiffel Tower (Figure 36). When it was unveiled in 1889, the 984-foot-tall landmark was the largest

Figure 36. The design of trees and plants facilitates two flows: water and stresses. Had there been two design objectives in the mind of Gustave Eiffel—mechanical strength and pumping water from ground to atmosphere—the Eiffel Tower might have been the Eiffel Tree.

man-made structure in the world, nearly twice the size of the previous record holder, the 555-foot Washington Monument. The American obelisk—like those built by the Egyptians—was an example of dry stone construction. The Washington Monument is held together by the weight of its stone. When he described his controversial design in 1888, Gustave Eiffel trumpeted its originality by saying it was "not Greek, not Gothic, not Renaissance because it will be built of iron. . . . The one certain thing is that it will be a work of great drama." The secret, of course, was its *natural* design.

Engineers have long been puzzled by Eiffel's design because conventional wisdom holds that such structures should narrow less rapidly as they rise so that the entire tower can support the largest weight. Gustave Eiffel's tower tapers like a truncated pyramid. This is because he recognized something that his contemporaries did not: His tower was so tall that it must be *uniformly* resistant to lateral bending due to the wind and axial compression due to weight, just like a tree. This apparent imperfection (deviation from the exponential) of the Eiffel Tower has been a puzzle until now. The genius of Eiffel's design was how it combined strength in compression (under the weight) near the base with strength in bending (subject to lateral wind) in the upper body of the tower.

Now we return to our theoretical tree and move aboveground. It is here that we get an even better appreciation of the splendid simplicity and efficiency of nature predicted by the constructal law. Recall the designs we saw in chapter 2, where river basins that pull water into the main channel look like the river delta that disperses that water, and the vascular system that carries blood from the heart looks like the one that returns it to the heart. We see the same thing in trees. Where roots collect water and move it up, the trunk disperses it up and out through the branches and leaves. Again, the tree can have an infinite number of shapes. We ask: Which is the better shape for the two flows that inhabit the trunk, with water moving up and stresses flowing to and from the ground from wind?

Not surprisingly, our theoretical design of the trunk produces the same shape we found for our root. This time, it is broad at the bottom and narrows as it rises because the higher we go, the less water we find as it is dispersed to the lower branches (Figure 37). This design extends to the branches, which, like the roots and trunk, should also resemble cones. Just as we find more small roots the lower in the ground we go—as the root system sucks

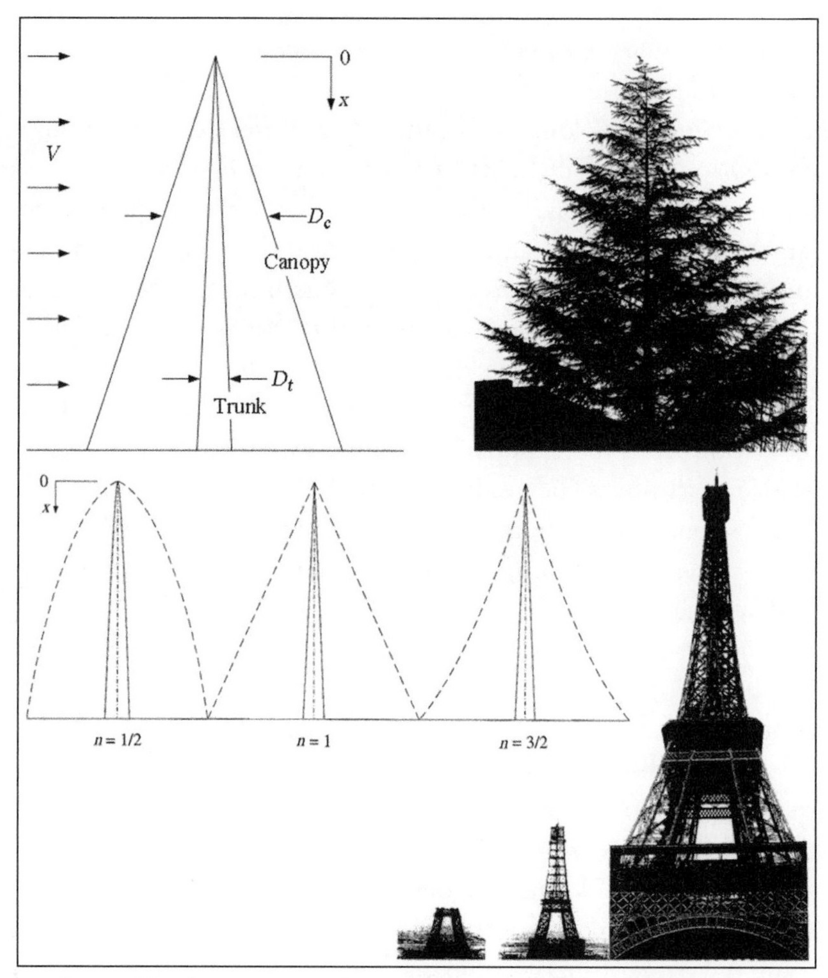

Figure 37. Three canopy shapes showing that the trunk shape should be nearly conical in all cases.

water up and then into larger channels—we find more numerous smaller branches the higher we go in the tree, because this is an efficient design for releasing water back into the air (Figure 38).

In the forest this image of a single design might seem puzzling. The canopies of different trees look very different—no one would confuse a weeping willow with a poplar or walnut. But this is akin to judging people by their hairstyles. When we strip away the foliage, we find that the trunks and branches of

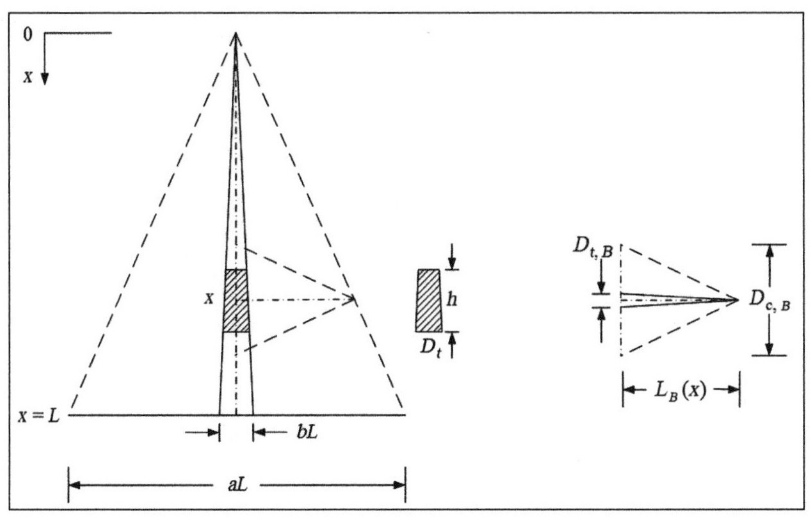

Figure 38. Constructal trunk: The constant ratio is h/x, where h is the vertical distance between two successive branches along the trunk and x is the distance from the level of the branch to the top of the trunk.

all trees and plants are tapered: They have essentially a conical shape. And, just as we found in the roots, the conical design that facilitates the flow of water is also the shape for handling the flow of stresses through the trunk and branches. Thus nature exhibits an elegant and uniform design for handling the multiple flows coursing through trees. These flow architectures are all consequences of the discovery of the conical shape of roots, trunks, and branches.

As we've seen in our discussions of river basins and animal locomotion, good design almost always involves scaling laws. Many of the great works of art and architecture are pleasing to the eye because they achieve a harmony born of beautiful balance. The design of trees and plants involves so many rules of proportion that you'd think they were dreamed up by a geometrist or artist. In fact, they remind us that the simplest solutions are often the best and once nature finds something that works, it runs with it.

A quick glance at trees reveals that they have more branches at the top than the bottom. This is not a random occurrence. A very specific design principle is at work. To understand this, let's imagine a tree that is 10 feet tall. Imagine that the trunk is a stack of many segments, one segment just tall enough to correspond to one branch that issues from the trunk, that is, the number of segments matches the number of branches; and at altitudes where branches are more numerous, the segments are shorter. If the first trunk segment is 1 foot long, we have established a trunk-to-tip ratio of 1 to 10. This then becomes the ratio that determines the distribution of branches along the entire trunk.

It follows that our second trunk segment that produces a branch should be 9/10 of a foot long in order to maintain this 1-to-10 ratio. The third segment should be 8.1/10 of a foot long and so on. The proportion is always the same, but the number of branches increases because the size of the trunk segment keeps getting smaller.

Before you rush out and apply this scaling law to a tree in your yard, remember, like all aspects of the constructal law, this predicts how trees *should* look. In fact, very few specimens reflect this precise design—though farm-raised Christmas trees come fairly close. In the real world, a host of environmental factors affect the branches. Strong winds, for example, bend them and shear them away. The lack of sunlight in a crowded forest, or on the north slope of a hill, imperils lower branches. Local environmental conditions are one of many factors that create variations in the shapes and structures of individual trees. The constructal law does not predict the particularities of design for every tree. It identifies, instead, the tendency of all the trees to evolve designs that facilitate flow. The result is the broad patterns we find in nature that are also affected by local conditions, idiosyncratic variations, etc.

Though it is rarely realized, this proportionality between the tree's branches and its height is the design direction that describes the evolution of the tree—the better flow pattern it generates. This proportionality also allows us to predict (in a few lines, with only pencil and paper) several famous empirical rules of tree design. One is the rule of Leonardo da Vinci, who in one of the folios of his amazing notebooks observed that the decrease in the cross-sectional area of the trunk as it tapers toward the tip is matched by the cross-sectional area of the lateral branch issuing from each trunk segment. This insight, which was based only on observations of the size of the cross sections, is correct. What Leonardo did not know was why this should be so. The reason, as the constructal law predicts, involves the flow of water and stresses through the tree.

Water moves up and out of trees through a series of interconnected strawlike tubes, called tracheids. To visualize this, imagine that you are building a tree with 100 straws representing the tracheids. When we reach the first branch of our tree, 10 of these straws split off to bring water to the first branch. This establishes a scaling law for our tree of 1 to 10 (different trees and plants have different proportions, but they are uniform for each one).

When we move to our second branch-producing trunk segment, we have 90 straws left in the trunk. In the next branch, 9 of our straws branch off into it. So it goes up the length of the tree, where 10 percent of our water-carrying straws angle out into each branch. As Leonardo observed, the cross-sectional area of the second trunk segment and the first branch equal the cross-sectional area of the first tree segment. The diameter of the third trunk segment and the second branch equal that of the second tree segment.

We were able to predict the proportionality between the size of the trunk segment allocated to one branch and the vertical length from the branch to the top of the tree ($h \sim x$; see Figure 38) by using theory to design a structure to facilitate the flow of water and the flow of stresses.

Another empirical rule of tree design that is a consequence of the constructal law is the Fibonacci sequence. This series of numbers, in which the two preceding numbers added together equal the next number (0, 1, 1, 2, 3, 5, 8, 13 . . .), has a long history. In botany, the Fibonacci sequence is encountered in the spiral arrangement of branches around the trunk. The constructal design of the tree demands this.

Here's why the spiral is necessary. For trees and plants to move water efficiently from the ground to the air, each of their branches should grow laterally into the space that is least affected by its neighbors. This is because all the branches are dumping water into the air, so the driest air will be the one farthest away from other water-emitting branches. Each branch grows and moves toward the space that contains the least humid air. The need to reduce interference between branches is a restatement of the constructal law, that is, the tendency to morph in order to have greater flow access for water from the ground to the air, that is, from the base of the trunk (the point) to the entire tree (the volume). The driest pockets that are available around the canopy can be visualized as the loops formed between two counterrotating spirals that rise to the tip on the outside of the canopy.

The two-spiral construction is not new. What the constructal law contributes is the prediction that the intersecting curves that form these loops must be spirals. Why must they be? Because the predicted proportionality between h and x (Figure 38) and the predicted conical canopy are the rules of how to draw spirals. In order to draw a spiral on a cone, one needs the cone and the rate at which the turns of the spiral get tighter as the spiral reaches the apex of the cone.

This is the power of theory. I am not a botanist, but the constructal law provides a principle through which I can predict a host of phenomena that botany currently studies. The constructal architecture of the tree canopy means that the total wood volume of any tree should be a predictable fraction of the total volume of the canopy ($\sim L^3$), where L is the trunk length (and the length scale of the whole structure). There should also be an optimal allocation of wood volume to leaf volume, such that larger trees must have more wood per unit volume than smaller trees. Furthermore, trees of the same size must have a larger wood volume fraction in areas with stronger winds. The total mass flow rate of one tree must be proportional to its length scale, L, or the canopy diameter viewed from above. All these features of vegetation design are in agreement with measurements across the board.

Let's return once more to the Duke forest. As we walk along its paths we encounter mind-bending diversity—a rich tapestry of tree canopies large and small. At first, the distribution of tree sizes seems a hodgepodge. We see small trees here and there, filling in gaps amid their larger cousins, finding niches where they can.

I suspect by now that you won't be surprised to learn that there is a method to this. Just as each tree is an individual pumping station whose design facilitates the flow of water from the ground to the air, the entire forest is a giant pumping station that mixes the number of large and small trees to achieve this on a grand scale (Figure 39). We have already seen that all trees are flow systems that must evolve in accordance with the constructal law. From this principle we have predicted their common design. But we don't stop there. Instead, we take a much larger leap in understanding by acknowledging that trees are inextricably linked to, and shaped by, all the other trees and forms of vegetation as well as the environment around them.

The constructal law proclaims that every building block and larger construct of any flow system should be guided by the universal tendency to generate patterns that increase flow access over time. Trees are components of a larger flow system, the forest, so their size and distribution should be determined by the forest's tendency to generate design that eases its own flow.

In a sense, nature is like a giant Russian nesting doll, but instead of dolls within dolls we see flow systems within flow systems. Each tree and each forest is a single entity, but all are fed and shaped by one another, just as each rivulet and stream is a morphing flow system that is also part of the evolving design of the larger river basin. Recall that in chapter 3, on animal locomotion, we saw that the shape and structure of the lungs and

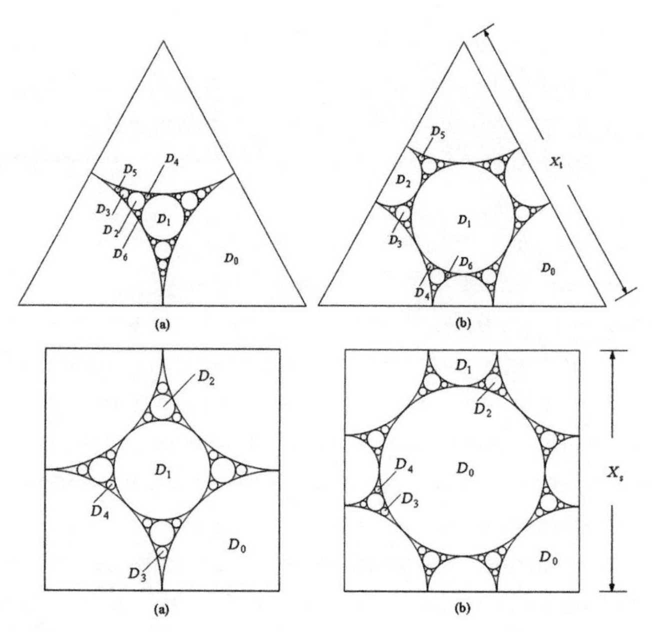

Figure 39. The design of the forest floor is a tapestry of tree canopies of many sizes: (a) algorithm-based generation of progressively smaller scales, and (b) more numerous larger canopies for greater global water flow access. The *D*s are the diameters of the canopies; X_t and X_s are the dimensions of the triangular and square elements.

heart manifest a two-track efficiency: They are good designs for oxygenating and pumping blood and are also efficiently designed components of the animal that move its body for the least amount of useful energy. Each tree works well on its own as an individual pumping station and also as a component of the much larger pumping station, the forest.

My colleagues and I explored the relationship between trees and the forest in a paper published in 2008 in the *Journal of Theoretical Biology.* We predicted that the best pumping design—that is, the best distribution of trees—would have a few big trees and many smaller ones because this was a good way to cover the entire forest floor with vegetation to move water. We tested this idea by creating a series of designs aimed at covering the entire area with trees as pumping stations. We started with the tallest trees because they move the most water—we predicted from the constructal law that each tree contributes to the global flow rate in proportion to its length scale. We gave these largest trees a rank of 1. Once we placed as many of the tallest trees as would fit onto our area, there were spaces between them. We filled these spaces with smaller trees that were given a rank of 2. This filled more space, but gaps remained. And so we continued the sequence, adding smaller and smaller trees until the entire area was in use. Our drawings revealed a pattern that might be amazing if it weren't entirely predictable: a hierarchical relationship among all the trees of varying sizes. Each of the areas had a few very large trees and larger numbers of progressively smaller

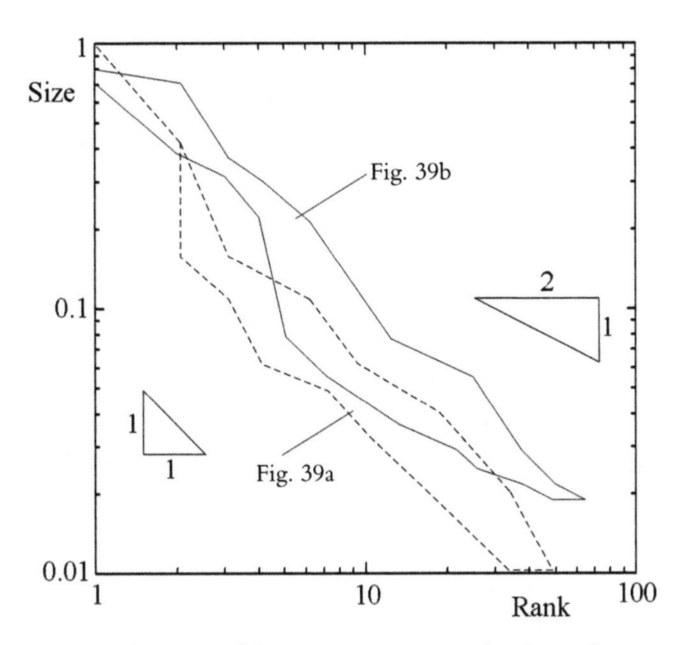

Figure 40. The sizes of the tree canopies on the forest floor versus the rank of all canopies according to size, as a summary of Figure 39. The largest canopy has rank 1 on the abscissa. All the tree canopies line up as a descending straight line with a slope between −1/2 and −1, which is known empirically. With the constructal law, this distribution is predicted.

ones. When we ranked the trees of our constructal forest by size, the log-log graph created straight lines with a consistent slope (between -1 and $-1/2$) (Figure 40). Testing our prediction in an actual forest, we found a similar hierarchical distribution of various-sized trees.

We showed that the trees and vegetation in the forest have a predictable hierarchy. Few large trees and many smaller ones is the blueprint for forest-floor design. Despite any Darwinian struggle for survival that might occur among all the forms of vegetation on the forest floor, the design is known in advance. This is also the physics basis for the emergence of hierarchy in nature that we will explore in the next chapter.

In summary, it is possible to place the emergence of vegetation architecture on a purely theoretical basis, from roots to forests. The key idea is the integrative view of design in nature as a physics phenomenon governed by the constructal law. In the big picture, each forest is a component of the global flow system—that includes rivers, oceans, and weather patterns—that reflects the universal tendency toward design generation and evolution to facilitate flow access. In this context, the forests are organs of the much larger global system. Similarly, each animal is an organ of the entire animal mass (essentially water mass) that flows across the continent.

This integrative approach reveals that the biosphere, atmosphere, and hydrosphere are not separate entities but interlocking environments that flow and design themselves together. From this idea we can predict their designs and the designs of everything that flows and moves.

This insight also challenges the Darwinian concept of winners and losers. In time, some species do flourish and others wither away. It even appears some thrive precisely because they are able to crowd out their "competitors," that is, their neighbors. The constructal law teaches us to see all flow systems as components of a single organism, the entire globe, which evolves its design to enhance its flow. They are not competing against each other but working together. The idea of winners and losers might make sense if evolution were a zero-sum game with no direction in time. But because flows morph to increase flow access for the whole, the whole becomes the winner.

We have shown that a constructal-law approach predicts the essential features of vegetation design. If the tape of evolution were rewound and restarted, and if vegetation design appeared again, the evolutionary process should consistently produce the same types of roots, trunks, and canopies—the forest-floor designs and scaling laws we see today. This tells us that there is a direction to evolution. It is not the story of random events but the unfolding saga of the emergence and evolution of design for better and better flow in time.

The constructal law places a physics principle behind Darwin's ideas about evolution. It tells us why certain changes are better than others and shows that those changes do not arise by accident but through the generation of design. The constructal law also expands our understanding of evolution, showing that the natural tendency of biological change is the same tendency that shapes the inanimate world.

As such, it provides scientific evidence for that soaring sense of oneness we feel when we walk in the woods. The ground, the trees, the air, and our own selves are indeed connected. Shaped by the same universal force, each sustains the whole in a grand symphony of creation.

CHAPTER 6

Why Hierarchy Reigns

Humanity's great fortune is that nature has shape, structure, configuration, pattern, rhythm, and similarity. It has rules and order: It is knowable, reliable, and, *on the whole,* predictable. From this stroke of luck, science was born and developed to the present day, where it is responsible for our well-being. If, for example, we couldn't count on the fact that water should boil at a certain temperature, that seeds should sprout into fruits and vegetables, that flow systems should evolve to move more easily—we'd live in a *Twilight Zone* world that is, quite frankly, unimaginable.

Just as we found different centers of mass for humans based on their areas of origin, trees have certain idiosyncratic variation encoded in their genes, so that a pine tree always looks like a pine tree and will never be confused with a weeping willow or a palm tree.

Since the dawn of science, we have worked to improve our understanding of the natural laws that govern our world. The foundational idea of the modern scientific method is that our findings must be verifiable and reproducible, which means that anyone anywhere should get the same result from the same experiment. Nowadays, even schoolchildren recognize the power of these principles when studying the mindless forces of nature. Less clear is the question of whether the intricate sets of laws and relationships that govern the natural world also shape the human-built landscape. The regulations established by various governments, much less the peculiar set of assumptions and expectations each of us holds as we engage all around us, often seem so capricious and contradictory that the idea of pattern and predictable evolution in human affairs can seem as hard to imagine as a physical world without them.

That, of course, hasn't stopped people from trying. A holy grail of Western thought has been to find principles as solid as those we find in science to explain complex social systems. Many of these efforts have involved the abuse, rather than use, of science. Karl Marx offered a "scientific" view of history that brought misery and death to the masses before it was discredited by experience. In the late nineteenth and early twentieth centuries, social Darwinists argued that the wealthier and more powerful people and races deserved to be on top because they were more fit than others. More recently, literary authors and social scientists have misused the concept of entropy (suggesting that all systems will tend toward disorder) and Heisenberg's uncertainty principle (popularized as the notion that ultimate knowledge is unknowable) to advance pessimistic and relativistic arguments about society.

In hindsight, we can see these for what they were: the misapplication of science to further ideology. Name-dropping is not science. And yet, like the mystics who had nothing more than intuition to support their sense of the oneness of nature, these efforts reflected a correct hunch. This is often how knowledge proceeds: Thinkers consider mountains of observations and piles of data, then try to find the pattern that links and explains them. At bottom people say, "I know something's going on here," but until now they haven't quite been able to put their finger on exactly what.

These efforts also reflect dissatisfaction with another wide current of thought, one that has separated people from the physical world and cast scientific knowledge as a tool we use to manipulate the world around us. This pragmatic "us" and "it" approach has paid huge dividends; the flourishing of civilization and technology has depended on our increasing understanding and control of nature, knowledge we use to create comforts and achieve progress. But it has left little room for the ancient inkling that we are, in fact, subject to these laws in profound ways. It cannot account for that sense that these laws push, prod, and shape us, guiding the choices we make about how we live, love, work, and play.

Now we can.

One of the most powerful insights born from the constructal law is that social systems are natural designs that emerge and evolve to facilitate the flow of the currents they represent on the landscape. This evolution has a direction in time, toward greater and greater access to move more mass (for example, people, goods, information) per unit of useful energy. The architecture and history of society, in fact, are not much different from the evolution of other complex (but simpler) flow architectures in the natural world: river basins and deltas, turbulence, blood vascularization, animal movement, respiration, dendritic solidification, and so on. All are throbbing, pulsing designs that evolve in time, that persist or perish based on their ability to facilitate flow access. Using the constructal law, we can, for the first time, predict the broad array of "pattern generation" phenomena that arises in geography, demography, communications, government, and economics.

This is a sharp break from conventional wisdom. It is also wholly unsurprising when we appreciate the fact that we sprang up from the world, were created by the same flows that generated all around us. The rise of humanity is not a radical break from the past but a chapter in the larger story of the Earth's long history of flow design and evolution. We are a part of nature; its oneness includes everything, even us. Our special gift is not the ability to act apart from nature but the ability to generate complex and evolving natural designs that allow us to move more mass faster, farther, with longer lifetimes, and more cheaply than other animal-mass flows.

As my Duke University colleague the sociologist Gilbert W. Merkx has written, this constructal perspective differs significantly from dominant approaches in the social sciences, which assume that structure is a given that sets the context for social action or transaction. Structures are seen as static and transactions as dynamic. To be fair, there is a literature about transitions between structures, but these transitions are anomalies, periods of structural breakdowns or "revolutions" leading to new periods of stable structure.

Constructal theory sees social structures (economies, governments, educational institutions, etc.) as flow systems that are dynamic, not static. Structure is not viewed as stable. Rather than being taken as given, the living flow structure is always in flux, ever evolving to provide better and better flow access. The evolution of flow structures reflects the interaction between time and the environment. The environment is important because it also evolves, altering the parameters within which flow occurs. Thus the environment is an essential dimension of any given flow structure. The environment, in turn, can be defined as a series of overlapping and interwoven flows that interact in space and time.

I call these environments of multiple, interwoven flows "tapestries." In nature, tapestries might be given labels such as "ecosystem" or "geomorphology," and in the human environment they might be called an economy or society. But they share the similarity that any single flow system within the tapestry is morphing its configuration to seek better paths in the context of other flows doing the same.

By proclaiming that societal flows emerge and evolve according to the same principle as all other natural flows, the constructal law challenges a long tradition dating back to Immanuel Kant. The tradition holds that there are two different realms of human knowledge, the natural and the human. Perhaps the most famous expression of this perspective is found in Max Weber's concept of *Verstehen,* or "sympathetic understanding." This idea states that the behavior of social actors is motivated by thought and culture, allowing an understanding of the reasons for behavior that is very different in character from explanations that describe behavior without reference to motivation.

If we grant that people, unlike drops of water, can think, then why should social flows come to resemble river networks? There are several reasons. The first is that social flows, too, are constrained by the physical world through which they move. So the movement of people will tend to be along paths of less and less resistance. Over time, transportation systems, like highway and railroad networks, develop treelike patterns much like river basins, responding to similar geographic challenges.

Another explanation of the similarity between natural and social flow systems is that the unique characteristics of each of the individuals that compose a system are irrelevant to the character of the flow architecture. No two leaves on an oak tree are identical, but they perform similar functions as members of the same tree system. Weber's concept of bureaucracy is premised on a similar assumption: that the rules of bureaucratic organization determine outcomes, not the unique characteristics of the individuals in a bureaucracy.

A third explanation is that individual motivations are canceled out in situations involving large numbers of people, a topic studied by the field of collective behavior. A final explanation is that while people's motivations may vary to some extent, most people, most of the time, are rational actors who aspire to decrease the costs and increase the benefits of their behavior. This is the basis of rational choice theory, which underlies modern economics. To the extent that people behave to maximize their benefits, they will construct and gravitate toward social networks that exhibit, or are believed to exhibit, efficiency.

The constructal law captures the broad tendency of social organizations to construct evolving flow systems that enable people and their goods to move more easily, more cheaply. This is not human desire. It is physics.

In previous chapters we have observed that almost all flow systems carry a current from a point to an area or from an area to a point. Although they are far more complex, human organizations are also area-to-point or point-to-area flow systems. Governments, corporations, religious groups, universities, sports teams, communication and transportation networks, cities, nations, etc., produce and transmit currents (goods, services, people, information, etc.) to an area through actual channels. Science, for example, generates *actual channels*—including scientific laws, schools, disciples, libraries, journals, and books—for the organization and spread of its current: knowledge. Religions create actual channels—including houses of worship, clergy, sacred books, etc.—for the flow of the doctrine to the faithful. Militaries also carve actual channels for the flow of strategies, materiel, soldiers, vehicles, etc.

Consider this streamlined description of the Ford Motor Company. Sedans and SUVs don't grow on trees. To create them, Ford needs raw materials to flow to its factory (a point) from the surrounding area. This involves multiple channels, including the lines of communication between the factory and suppliers—"Send us ten tons of steel and a million tires"—and the various transport routes (roads, train tracks, air transport routes) generated to ferry those materials to its factories, channels that are now strategically placed around the world to allow the company to increase efficiency. At the factory, supervisors use channels of communication to direct workers and machines on the assembly lines. Once the vehicles are manufactured, they are sent out into the world (an area) through actual channels that bring them to the dealerships that use their own channels of communication (advertising, word of mouth, etc.) to reach customers. All these channels evolve in time. Some become larger, others smaller. But the changes that stick are those that allow the flow system to persist in time.

This is the basic design of many businesses. By design, we mean something very specific—the *actual drawing* that flow systems create over time. We are not talking symbolically here, using analogy or metaphor. The constructal law is not an abstract theory but pure physics, observable nature, and unifying principle. It predicts the movement of physical entities over the globe, the flow of things we can see, hear, feel, taste, and touch. It is the black line on the white paper, the road on the map. The drawing is not a visual suggestion, but the design itself. A lightning bolt *is* the tree-shaped architecture that evolves in a flash to move electricity from a cloud to the ground. A river basin *is* the collection of waterways whose treelike structure moves water from the area to the river's mouth. Ford *is* the vast global structure of channels and interstices through which currents of materials, products, and information flow. If these currents were to stop flowing, the factories would be dead buildings.

To appreciate this point, we need to clarify our language. So far we have used the phrase "treelike structure" to describe the, well, treelike structures that abound in nature. It is a vivid image that paints a pretty accurate picture. And yet, as my work on the constructal law evolved through the years, I found that this verbal symbol—this flow system for conveying information about the law to others—does not provide access to all the meanings running through it. I have resisted until now using the better word in this book because it is a mouthful—"vascularization."

This word is an improvement, a better channel for communication, because it captures the central idea of the interdependency of life. Where a tree suggests the connection between point and area or volume, "vascularization" also contains the pivotal idea of *life-giving flow* and of a body (volume, area) filled with life. It reminds us that design arises in order to spread often-nourishing currents across an area or throughout a volume. The most familiar template for this is the vascular structure of our circulatory system, which delivers life-giving blood to all the cells in our bodies. Similarly, for a business to persist in time it must deliver life-giving ideas, materials, and goods to all its workers and customers.

This active, throbbing sense of design was contained in my original formulation of the constructal law sixteen years ago: "For a finite-size flow system to *persist* in time (*to live*), its configuration must evolve in such a way that it provides easier access to the currents that flow through it." It also harkens back to the seemingly radical definition I offered in the introduction for what it means to be alive. There I said that everything that moves and morphs in order to flow and persist is alive. We can refine this now to say that everything evolves in order to provide greater access to the life-sustaining currents that run through its vasculature. When nothing flows through our bodies, we are dead. When the water stops moving through the river basin, it too is dead; when material and information stop flying to and from a business, the business withers and dies. And so it goes, with everything.

To show that human organizations are governed by this principle of physics, we should find two features: The patterns of their channels should have a vascular shape and structure as we observe in other point-to-area flow systems, and those patterns should evolve in time to provide greater flow access.

This is what we find. To see how, we have to introduce a cornerstone characteristic of natural design—hierarchy. Although it has received a bad rap as a symbol of inequality, hierarchy is essential to good design. Instead of providing advantages to one entity to the detriment of another, it arises naturally because it benefits the entire flow system, whether it is all the water in a river basin or all the people in a society.

Hierarchy evolves because good flow often involves multi-scale architectures—that is, channels of varying sizes. On our commute we travel through many channels: highways, avenues, boulevards, side streets, and the path from our front door to the garage or curb. Even though highways handle many more cars and allow higher speeds, they are not more important than smaller avenues or even cul-de-sacs. All are necessary to spread the current that runs through it (us) to every destination, which means to provide access to the whole area. The few large feed the many small, and vice versa. An efficient transportation system strikes the right balance among all these components, just as other flow systems—including the distribution of people, wealth, knowledge, and education in society—should among their multi-scale channels. The fact that hierarchy arises across the board offers further proof that the constructal law is a universal principle of physics.

As we have seen time and again, the constructal law was just waiting to be discovered. Its manifestations are so obvious and ubiquitous that we have danced around it for centuries—the hunches of scientists, the metaphors of poets and mystics, and everyday language (for example, "the tree of life," "go with the flow," "the path of least resistance," "if you can't beat 'em, join 'em," "all roads lead to Rome") suggested a phenomenon they could not quite capture. So it is with hierarchy. About twenty-four hundred years ago, Aristotle coined a famous phrase that also hinted at it—*the one, the few, and the many.* While the Greek philosopher was defining distinct political systems, he also suggested the hierarchical structure of design in nature. In our research group, this most essential feature is conveyed and taught by Lorente's vision of *few large and many small.* Lorente's insight is much closer to the truth than Aristotle's because hierarchy means not only numbers but also sizes. "Few large and many small" is the concise name for what others describe as the emergence of "complex" design and "hierarchy."

Start with the river basin. There is always *one* main channel—the Mississippi, the Danube, the Seine. It is the widest, fastest-moving component of the flow system. It is fed chiefly by a *few* large streams. These, in turn, are sustained by *many* small tributaries and rivulets in an immense system that moves water from the entire plain to the river mouth. Or consider the tree. It has one main channel, the trunk, a few main branches, and many roots, stems, and leaves. All are necessary for the efficient flow of water from the ground to the air.

This same hierarchical structure occurs in our bodies. As oxygenated blood leaves the heart it enters the one large main channel—the Danube of the circulatory system—the aorta, which continuously branches into a few arteries and many capillaries. Now take a deep breath. Feel the air whoosh down the respiratory system's single main channel, the trachea. From there it enters the lungs, where it fills a few air passages before saturating the many tiny alveoli. The structure of lightning bolts, snowflakes, and lava flows, of forests, coral colonies, anthills, even clumps of dust—*every* multiscale flow system exhibits a hierarchical structure composed of a few large channels and many smaller ones.

And so we find the same design in human organizations. Almost every government has one leader—the chieftain, king, sultan, president, prime minister, governor, or mayor—who, like the main river channel, must handle the most important flow of information and authority. He or she is assisted by a few streams of top advisers, who themselves work with and oversee the many individuals who form the bureaucracy. This same hierarchy, which is often described as "vertical integration" in the business world, defines the structure of most corporations (one CEO, a few top managers, many workers), universities (one rector or president, a few provosts and vice presidents, more deans, even more department heads, and many more professors, teaching assistants, and students) and sports teams (one head coach, a few assistant coaches, many players).

The military, of course, is the classic example and main laboratory of this hierarchical chain of command. The Romans, for example, were able to expand their empire in no small part because of the hierarchical design of their military. Today, the United States of America has the most powerful military in the world, with a complex hierarchy spread across its major branches. People have written entire books analyzing this structure. Using the constructal law, we can predict that its structure should have a few large channels and many smaller ones, a fact confirmed by this very broad overview of the chain of command of the U.S. Army.

The president is the single commander in chief, responsible for all major decisions. Below him is the secretary of defense, who instructs the United States Central Command, which develops and implements strategy with commanders from each of the four branches.

Lieutenant Colonel Brian De Toy, director of the Defense & Strategic Studies Program at the U.S. Military Academy, told me that while there is plenty of variation in military formations, there are also clear patterns. By and large, the largest group is the army, which is commanded by a lieutenant general. The maneuver/fighting elements of the army are usually composed of two or more corps that are also typically commanded by lieutenant generals. Each corps is composed of two to five divisions, each commanded by a major general. This pattern continues as we proceed down the chain of command. Each division has three or four brigades; each brigade has three or more battalions; each battalion has three to five companies; each company has three to four platoons; each platoon usually has three to four squads; and each squad is divided into two teams.

What jumps out at us is that the hierarchical design has a predictable construction pattern very similar to what we find in the evolution of the river basin—essentially a rule of quadrupling, with each group composed of about four subunits.

Social systems generate hierarchical structure for the same basic reasons that other flow systems do: because point-to-area and area-to-point move their currents more efficiently with it than without it. Note that the constructal law makes no value judgments. Enlightened democracies and rigid dictatorships both display hierarchy, as do well and poorly run companies. What the constructal law predicts is that hierarchy should emerge naturally as a result of the tendency of moving things to generate designs that facilitate flow access. The constructal law also predicts that the rigid hierarchy will give way in time to a freely morphing hierarchy. This is why dictatorships are relatively short-lived and democracies have staying power.

Until now we've been looking at one snapshot—an entire river basin, the circulatory system, a tree, university, or corporation. As predicted, we found a hierarchical structure of multiscale channels defined by the few large and the many small. When we narrow our focus, we make the same observation. In the circulatory system, for example, small arteries are main channels for the network of even smaller arteries and capillaries they nourish.

Naturally, we find the same phenomenon in human organizations. As we saw in our sketch of the U.S. Army, the president is the main channel. He works with the secretary of defense, generals, and others to transmit orders to a hierarchically designed network of subordinates of various ranks. Except for the lowest soldiers, all of them have superiors, yet all have their own turf. Captains, for example, are the main channels of authority for each company; lieutenants serve the same function for members of their platoons. The Catholic Church also has an immense, complex hierarchy, from the pope to cardinals, archbishops, and on down. But at the parish level, the local priest is the main channel at the top of the hierarchy that includes monks, nuns, altar boys, and worshippers. When we look at an entire business, the CEO is the main channel. But as we go down the level of authority, from senior and middle managers to the foreman on the factory floor, these tributaries serve as main channels for the streams they feed.

We find hierarchy at almost every scale because flow architectures evolve in accordance with the constructal law. They do not generate hierarchy at the end of the day but at every step along the way. As soon as the seeping water starts coalescing to form rivulets, a hierarchical structure of a few large and many small channels emerges, which has evolved into the massive river basins that cover the globe. Similarly, the internal structure of multicellular organisms has a hierarchical structure, albeit one that is far less complex than that found in animals such as humans.

Flow systems continuously generate design for easier flow with hierarchy. This can be hard to see sometimes, in part because flow systems are always evolving, ever morphing toward better hierarchical configurations. I remember hearing an author on the radio claim that the early Catholic Church was not hierarchical. This statement appeared true to him because he was comparing the primitive group of churches to the highly organized structure that emerged, especially after Emperor Constantine embraced Christianity and advanced it throughout the Western Roman Empire during the fourth century. The early church had a more localized, less elaborate hierarchy, but in each community of Christians there were leaders, and certainly, very early on, St. Paul was the main channel for the dissemination of the doctrine.

This highlights the false assumption that because one flow system is younger and less evolved than another, it lacks hierarchy. As we have seen, the river basin is a more complex flow system than the one created by tiny rivulets, but both exhibit hierarchical design. Similarly, the government of the United States has a much more intricate hierarchical design than the network of tribal chiefs and warlords who hold sway over large swaths of Afghanistan or Pakistan. But even there, hierarchy rules.

This underscores a key point, and the fresh perspective, provided by the constructal law. The vascular design of the evolving architectures depends on the size of the flow system. Large structures have more levels of branching, that is, designs that are more complex, than smaller ones. In the opposite direction, smaller flow systems generate simpler tree shapes with fewer levels of branching.

While all multiscale, point-to-area flow systems generate easier flowing configurations that have hierarchy, they do not simply reiterate the simplest design into ever more complex patterns; the smallest detail is not simply a miniature version of the largest drawing. Just as one size never fits all, neither does one design. Flow systems generate *just enough* complexity for the size of the territory they bathe with current. They create architectures that work for them. The complexity of each architecture is modest, finite. The phenomenon of design in nature is not one where complexity increases in time. Each flow system evolves to acquire the right level (kind) of complexity to flow, to live. This is one reason why "simple" life-forms have persisted for billions of years even as other, more complex organisms have emerged. The phenomenon of design in nature also covers the rare cases of some fleas and tapeworms that have become simpler over time. The tendency in nature is not toward greater complexity but better flow access globally. This direction often gets lost because many natural flow systems become larger in time, and their finite complexity increases.

If we zero in on the subvolume between two alveoli in the lung, we do not rediscover the structure that resembles the human lung with its twenty-three levels of branching. Instead, we find the soft and wet tissue with diffusion, that is, with no distinct currents. Similarly, a local parish may have a hierarchical structure, but it is not a miniature version of the hierarchy we find in the entire Catholic Church. What we find is an evolving architecture of channels that handles its flows efficiently.

Let's take a closer look at the evolutionary emergence of hierarchy in social systems by examining a flow system that is particularly close to my heart—scientific knowledge. Science is what we can say about nature; it is our knowledge of how things are in nature (around us and in us) and how they work. In its rawest form, science is a collection of many observations. The sun is in the sky. It feels warm. It disappears at night. The moon appears. The temperature drops. If science were only a collection of such statements, it would not be very useful. In time, scientists have organized and improved this deluge of information in the same way that a river basin has evolved: toward configurations (links, connections) that coalesce (condense) the flows and provide faster access for the flow of information.

For example, prehistoric humans knew very little about how things fall. They had no concept of gravity. Through experience, hunters learned to throw a rock or a spear so that it would strike their prey. Each generation transmitted this knowledge to the next, a flow of information that was effective but relatively inefficient as it required face-to-face training and firsthand experience as individuals developed a feel for the work. Over time, science progressed as people accumulated bulky measurements. Greek and Roman soldiers developed basic formulas for determining how far away they should stand from their adversaries to hit them with their arrows or the payloads of their catapults. This represented an advance in knowledge. The early sense of intuition possessed by hunters (the rivulets and first main channel) was replaced by these calculations (the new main channel for the flow of information of how things fall). The armies that possessed this knowledge and that had created channels so that it could flow to their soldiers in the field had an advantage over opponents who lacked it.

Then, Galileo stated the principle that rendered all those measurements unnecessary: Objects fall faster and faster downward at a predictable rate. He gave the world a single formula that allowed us to replace that collection of discrete calculations. Thanks to him, we could predict the speed at which any object falls. His discovery spread through various channels—books, disciples, etc. Unlike many other scientific breakthroughs, including his defense of Copernicus, Galileo's discovery did not meet much resistance from competing ideas or entrenched dogma. Because it enhanced the flow of knowledge, it became a new main channel in the hierarchy of science. Indeed, all the great discoveries, from Newton's laws of motion to the laws of thermodynamics, didn't just tell us something new, they also organized and streamlined our knowledge. They replaced bulky measurement with principles that serve as new main channels in the hierarchical flow of knowledge about how things should be in nature. They didn't just rewrite our science books, they made them thinner, easier to teach and to learn from, enhancing the flow of information from those who possessed it to those who wanted to have it. Today, the constructal law is uniting a host of seemingly far-flung phenomena—design and evolution of the inanimate and the animate—through a single principle of physics.

We "know more" because of this evolution of flow design in

time, not because our brains are getting bigger. We keep up with the steady flow of new information through a process of simplification by replacement: In time, and stepwise, empirical information (such as measurements, data, and complex empirical models) are replaced by much smaller summarizing statements (such as concepts, formulas, constitutive relations, principles, and laws). Empirical facts (observations) are extremely numerous, like the hill slopes of a river basin. The laws are the extremely few big rivers. A hierarchy of statements emerges naturally because it facilitates the flow of information. It is an expression of the never-ending struggle of all flow systems to design and redesign themselves. As the constructal law predicts, better-flowing configurations continually replace existing configurations.

In a river basin, the marsh and the incipient rivulets that form after rain are akin to the relatively unorganized, raw volume of scientific data. Over time, they continuously generate ever more complex—and better—flow channels of rivulets, streams, and tributaries. In science, the evolving channels are language, subjects, laws, schools, disciples, libraries, journals, and books. The same thing happens in all social systems. Civilization is the story of better and better flow access. The evolving design of politics, economics, technology, and all the rest have created channels that improve the movement of people, goods, and ideas. It's true that every change has not enhanced flow access; bad ideas are inevitable, and entrenched powers often try their best to defend their limited interests. In the short term, evolution looks like a jagged line. But in the long run, there is a clear direction in time: The currents that persist are those that facilitate movement. This is progress.

Scientists are constantly making new observations, just as the skies regularly pour rain on the ground. Both flow systems are *preexisting* and *evolving*. The first part of this statement is easier to appreciate than the second when we are looking at a river basin, because its design has emerged over millions of years. During that long period of time, each of its channels has constantly configured and reconfigured itself, finding better and better ways to move the water within its specific geographic area.

This time element explains why we don't see new large rivers on the landscape. The river basin is still evolving but it is also entrenched. Now, if the worst fears about global warming were to come true or some other cataclysm were to dramatically alter the current system—if the middle of America turned into a desert and the Sahara became a floodplain—two things would happen. In the worst-case scenario (for America, anyway), the Mississippi would dry up and a new, hierarchal flow design would arise in what had been the African desert.

Science is a much younger and far more complex flow system than the river basin. So our drawing of scientific knowledge is changing far more quickly. The falling raindrops of this field (raw data, observations, creative minds) spark the visible creation of better channels (new laws) to handle the current (knowledge and ultimately the movement of humanity). To the extent that these new channels are able to handle the flow of all the observations that have come before and those still to be made, they will become deeper, more entrenched.

As an aside, I'll add that this is one reason that scientific ideas appear to take a long time to take hold. As Max Planck observed, "A new scientific truth does not triumph by convincing its opponents and making them see the light, but rather because its opponents eventually die, and a new generation grows up that is familiar with it." For lack of better laws, researchers try to squeeze new data into the old channels. They try because they depend on and benefit from the existing design, which is called the establishment. The establishment fights back, but after enough funerals it becomes a new structure, a new establishment. Truly original ideas break this mode and replace the existing structure with one that flows better. This is what the constructal law predicts for the evolution of science.

This view underscores another seminal aspect of hierarchy:

the interconnectedness and interdependence of every component of the flow system. In the bigger picture, the river basin uses its rivulets, streams, and main channels to move water from the plain to the river mouth just as surely as our respiratory system uses tiny alveoli, bronchial tubes, and the trachea to bathe our lungs with oxygen. CEOs use managers and workers to create their company's products and spread them to their distributors and customers, just as presidents, prime ministers, and dictators use advisers and countless bureaucrats to develop and spread their policies across the land. The main channels may facilitate wider, faster flows than the smaller ones, but all are necessary to bathe the entire area with the current. Hierarchy emerges because all flow systems use the right combination of components of varying sizes to efficiently move the currents that flow on the same, finite territory.

This finding leads to another insight that debunks conventional wisdom: Hierarchy arises because it is good for *every* component of the global flow system. The big need the small just as surely as the small need the big. The individual sustains the crowd—and vice versa. The big river sustains the many tiny streams of the river basin, just as those tiny streams feed the river basin. Citizens (the rivulets of politics) sustain the governments that serve them; workers (the rivulets of business) sustain the companies that employ and, in turn, sustain them. The urge to organize is selfish.

It is the integrative aspect of design—the balance that naturally emerges among all its flow components—that is one of the most revolutionary insights offered by the constructal law. While the prevailing Darwinian model of evolution makes some room for the idea of cooperation, it is based chiefly on the idea of struggle among individuals—the "good me" against my bad neighbors and society. Organisms compete with one another for scarce resources; we compete with the environment, etc. It is, largely, a tale of winners and losers.

The constructal law, however, reveals that the movement toward harmony, toward flowing together and in balance, is the central tendency of design in nature. Recall our discussion in chapter 2 about the predictable ratio between the number of daughter and mother streams in a river basin and the ratio between the mother and daughter air tubes in the lung. We showed that these relationships and the many scaling laws in nature have emerged because they are efficient designs for global flow access. Nobody is commanding these various waterways or blood vessels to act in concert, but they do, each part serving the flow performance of the whole. Similarly, human organizations thrive when they find the right balance of flows. Employees can strive to earn a particular position, and their success or failure at this can have personal results. But for the outfit to run efficiently, it must find a combination of workers at every level for its size at a given time that changes during periods of boom or bust. This is why hard-working, talented people often lose their jobs. Their best efforts no longer match the changing needs of the bigger system.

In a larger context we see that black markets and smuggling emerge where the official channels do not facilitate everything that flows. Under communism, for example, there were thriving black markets for almost every commodity because governments imposed artificial designs and constraints. In the United States of America today there are few black markets, but those that have emerged (for example, for illegal immigrants or illegal drugs) are in response to laws that hinder flow access.

So far we have seen how this works in relatively self-contained flow systems such as trees and river basins. In both cases it was easy to see that the flow system tends to strike the right balance among its multiscale channels and their interstices. Now we will widen our field of view to see how this predictable harmony of hierarchy occurs in even larger flow systems. The constructal law allows us to discover that the same design governs the relationship among all the seemingly independent flow systems that form various flow networks. I recognize that this is a fairly radical point—but it is also an obvious one when we take a holistic view.

To achieve this perspective, let's pretend we're on a space station orbiting the Earth. We see below a vast map covered with innumerable flow systems, each of which is evolving in time to provide easier access for its currents. Everything is connected to everything else as part of a global system that changes in order to flow better and better. Every river basin is part of the global system that includes ocean currents and global weather patterns and that tends to move toward equilibrium of all the heat and moisture on the planet. Every business is a component of local, national, and world economies. These larger systems are, of course, evolving. Because the global system to which the river basin belongs is less complex and far older, it displays a higher level of integration than the economy. Its channels have had more time to find the right places and combinations and so they are more entrenched. In time, we should also expect the channels of the economy to become more integrated, to provide easier and easier flow. They are, and we call this globalization.

Globalization is as old as civilization itself. It is an evolving design that began when the first people migrated and when the first individuals and tribes began trading with one another (both goods and ideas). Today's world is just the latest chapter in this long story as technology has made it cheaper and easier for the entire globe to put the right channels (people, goods, ideas) in the right places.

I explored this phenomenon—the fact that seemingly independent flow systems are connected to larger systems that also display hierarchy—when our research group applied the constructal law to the design of vegetation (see chapter 5). After using it to predict the design of trees, plants, and roots, we figured: Since the forest is a giant pumping station for the movement of water from the ground to the air, it should also manifest a hierarchical design that maximizes the ground-to-air fluid flow access. We should find a few big trees and larger numbers of trees, grass, moss, etc., at progressively smaller scales. We then confirmed our theoretical findings in the real world.

This dovetailed with the findings of an earlier book in which my colleagues and I had used the constructal law to predict the same striking pattern in a far more complex flow system: the size and distribution of human settlements. We began again with pure theory—the idea that the flow system of demography must bathe the area (the continent) with people, goods, and services. To do this efficiently, it should have settlements of varying sizes that are proportional to one another and to the entire area.

Human settlements are a flow system for the movement of people and their goods, ideas, and so on. Like other natural flow phenomena, their evolution hinges on finding less travel time and cost for all these things. In the simplest description, civilization is the name for the coexistence of farmland with the market. Those who live on the area exchange farm products (and other goods, services, and information) with those who manufacture products and deliver services in compact places—first hamlets, then villages, then small towns, and finally cities.

Evolution depends on technology—significantly larger settlements can emerge only as technology allows people to minimize the travel time over greater distances. The allocation of land to human concentration also predicts that there should be proportionality between the number of people in the larger settlements and the surrounding areas. This balanced design should emerge at every step of the evolution of human settlements.

With this idea in mind, we created a series of ever larger constructs over an area with populations that remained proportional. As we added more settlements, hierarchy developed in two ways: Areas coalesced, from the smallest element to the first construct, and the population developed concentrations from farmers on several plots to traders at several points and finally to one trading point, which was, perhaps, a small town. The end of this sequence occurred when the constructal area matched the size of the available area. At this uppermost level of assembly, the number of large cities was one or two.

When plotted logarithmically, the size-versus-rank distribution of all the settlements formed the same pattern—a straight line with a descending slope—that we predicted for trees in the forest. This theory also predicted that this line should shift upward while remaining parallel to its previous position because of technology evolution: In time, each unit area sustains the living (the movement) of more and more inhabitants.

Next, we tested our finding in the real world by plotting the size and distribution of cities in Europe since 1600. The data confirmed our prediction: The populations of the various cities were always proportional, and they always generated a straight

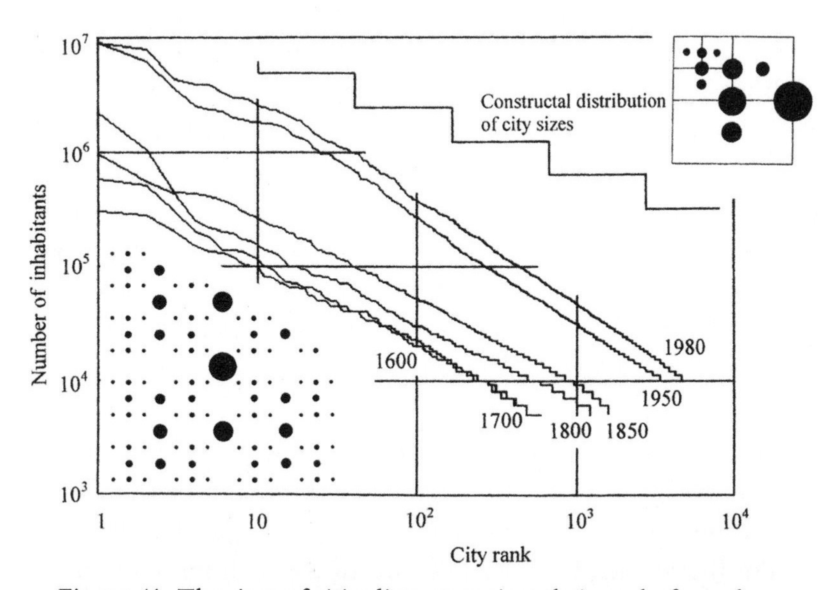

Figure 41. The sizes of cities line up against their rank, from the largest (1 on the abscissa) to the smallest. They form a straight line, which slides upward in time as technology, standard of living, and GNP rise. The curves show the city sizes versus city rank in Europe from 1600 to 1980. The stepped line shows the distribution predicted with the constructal law. The detail in the upper-right corner shows that the size of every city matches the size of the area on which the city participates in area-to-point flows. The detail in the lower-left corner shows the predicted distribution of multiscale human settlements on the map (few large and many small). This pattern is obtained after deleting the construction lines from the upper-right detail.

line with a descending slope on our graph. Other researchers who have recorded the size and distribution of human settlements around the world have reinforced this finding (Figure 41).

Nevertheless, it is important to note that this construction of compounding areas is not how human settlements form, grow, and get connected on the landscape in time. The compounding of areas (Figure 41, upper-right corner) is about how the tapestry takes concrete shape and structure in the mind, once we discover that those who live on the area must be proportional to the number of those who live in the settlement (because both numbers must be proportional to the same area).

On the ground, and in history, settlements become denser and each settlement grows in time. But they do not grow at the same rate. Some grow faster, and when they become large enough, they incorporate their small neighbors. And, from among those that are becoming larger, an even smaller number—a special very few—will grow even faster, in proportion with the growth of the population on the larger and larger areas that appear to be allocated to the special few. And so on, all the way to the single megacity on the map.

Because the emerging map of demography varies from country to country, the evolution to multiscale design can seem random. Luckily for the theoretician, this is not the puzzle. The real issue is the pattern, the broad features, performance, and evolution of the whole. This issue unifies all the seemingly unrelated counties, countries, and continents and it is predicted fully with the pattern shown in Figure 41. This is the power of the purely mental viewing and the constructal law.

I wasn't aware of the breadth of this discovery until 2005, when I was presenting this theory at the Sun Valley Writers' Conference. After my presentation, J. S. Adams, an engineer who worked in the city planning department, came up to me and said that I had predicted something called Zipf's law. This was news to me. Zipf what? Who?

I learned that George Kingsley Zipf was a Harvard linguist

whose studies included how often particular words appear in the English language. In a paper published in 1935, Zipf reported that the occurrence of any word is inversely proportional to its rank in the frequency table. That is, the most common word appears twice as frequently as the second two most common words, which appear twice as frequently as the next four most common words. When plotted on a log-log graph, this alignment of rank versus frequency creates a descending line.

Zipf's work was refined by two scholars from Brown University, Henry Kučera and W. Nelson Francis, who performed an extensive computational study of English usage. They found that the most common word, "the," accounted for 7 percent of all the words in the wide array of texts they studied. The next two most common words, "to" and "of," each represented about 3 percent of the words used. As they continued their rank-versus-frequency study, they found that the third most frequent group of words was larger than the second group. Examining the ever-expanding groups of the fourth, fifth, sixth, etc., most common words, they corroborated Zipf's line. They found that about 135 words account for half of all the words used in English. Think about it: We don't say "ameliorate" or "egregious" very often.

No one would argue that "to" and "of" have outcompeted "ameliorate" and "egregious." They have not emerged victorious in a Darwinian struggle, a dictionary war. The truth is that a hierarchy of words has emerged naturally. This becomes clear when we recognize that in written and spoken communication, words and sentences are the channels that carry the currents that represent the thoughts and feelings we wish to express. In order to spread this current efficiently, a hierarchy of channels has evolved of large channels ("to," "of") and small channels ("egregious," "ameliorate"), all of which are necessary for the flow of information, and for our own flow (movement) on the globe.

Zipf titled his final book *Human Behavior and the Principle of Least Effort: An Introduction to Human Ecology*, and so we find yet another correct hunch. Other researchers have developed ad hoc principles to describe naturally emerging hierarchy in a variety of areas, including the distribution of wealth in society (Pareto's principle), the frequency with which digits occur (Benford's law), and the flow of scholarly publications (Lotka's law). I have not confirmed their results, but I might have predicted them.

To take a more recent example, in a July 8, 2010, column in the *New York Times*, "The Medium Is the Medium," David Brooks echoed the conventional wisdom that the Internet "smashes" hierarchy. At first glance, this insight seems obvious. Everyone who has watched the nation's mighty (and not-so-mighty) newspapers suffer a million blog bytes knows the mainstream media is getting pummeled, beaten, and smashed by the World Wide Web. Innumerable folks with laptops and videocams are supplanting the towering edifices of modern journalism.

The hierarchy cannot hold; mere anarchy is loosed upon the world.

This view casts new media and traditional media as opposing forces, two fighters battling over the same piece of turf. In fact, the constructal law reveals that they are actually complementary channels in the global flow system spreading information over an area (populated by people who receive and use that information to move more easily across the landscape). Just as the river basin carries water from the ground to the river mouth, the Internet, newspapers, and other forms of media are channels for information.

The Internet is swelling as traditional channels shrink because it can facilitate heavier flows (more efficiently, to boot) than the structure generated by traditional media. One view of this intricate design shows us that the Internet is the new big river basin—the Mississippi of the Information Age—fed by a few large streams (such as YouTube, Facebook, and Brooks's *New York Times*) and many smaller ones, including the millions of blogs and personal Web sites and billions of e-mails and instant messages sent each day. Like the design of science, the design of the Internet is evolving before our eyes to facilitate the flow of information. Tracking studies by Complete, a web analytics firm, show that the top ten Web sites accounted for 31 percent of U.S. page views in 2001, 40 percent in 2006, and about 75 percent in 2010.

All make up the evolving design of the global flow system for information. The rise of the Internet does not reflect the demise of hierarchy but its evolution and constructal design. It is larger and mightier (better flowing) than what came before. This is the same reason horse-drawn carriages were supplanted by automobiles but not replaced. Both could get you where you wanted to go, but one was much better than the other, and the two together are even better. That transition was not a radical break but a continuum in the evolution of better and better flow for the movement of people and goods predicted by the constructal law. This hierarchical structure emerges naturally because it facilitates the flow of information. And, in turn, the flow of information facilitates our own movement on the globe. This design change is analogous to the emergence of vision (the eye), which stepped up animal movement from groping in the dark to "guided" locomotion (see chapter 9).

Using the constructal law we say: The old hierarchy cannot hold; a better hierarchy is loosed upon the world.

Finally, the discovery that social systems emerge and evolve just like other natural phenomena raises what we might call the constructal paradox. Because human beings have consciousness, it is relatively easy to understand how we can organize ourselves into efficient flow systems. Unlike river basins and lightning bolts, we are smart and capable, and we can learn from the past in order to exert control over the future. When we look at the evolution of governments, corporations, religious organizations, etc., we see mankind's intelligence at work—willful, calculated, purposeful.

This can make it hard to appreciate the fact that our actions are guided by a natural tendency. For all our thinking and debating, our long record of achievement, our torturous history of conflict, we have generated natural designs. And our legacy for having done this is the same as the legacy of the river basins: We moved more mass than we would have without such designs.

In chapter 3 we showed that if we reran Stephen Jay Gould's tape of life and started it all over again, many things would change but not the constructal design of animal movement. Similarly, if we rewound the tape of human history and hit the record button anew, the movie would likely have different scenes and players than the ones we know, but the hierarchical design of our social systems would remain.

Here's one way to think about it. People have created innumerable currents through time: transcendent ideas and can openers; miraculous medicines and Frisbees; cinema and basketball; air-conditioning, indoor plumbing, and pool tables. As useful as they all may be, none was inevitable. If history had taken a different turn, we might find other currents. But those currents would facilitate our flow through multiscale channels in order to reach all the people they serve, which means to enable all the people to move more easily on the globe.

That our social systems have the same evolving design as other natural phenomena alerts us to the fact that forces far larger than ourselves are in play. It shows that our movement on the landscape is governed by the same principle as movement all around us.

What our history makes clear is that human organizations are evolving like other flow designs because they are not separate from nature but a part of it. A government must bathe a nation with rules and policies; the Internet must spread knowledge around the world; and a corporation must deliver goods and services to its customers. All generate vascular designs with hierarchy, all go with the flow.

CHAPTER 7

The Fast and Long Meets the Slow and Short

Most people don't like airports. Worse than the overpriced sandwiches and "unforeseen delays due to weather conditions" is the feeling of being stuck, boxed in among the confining crowds.

But the Hartsfield-Jackson Atlanta International Airport is different. It may be one of the world's busiest hubs, serving almost 90 million passengers each year, but when we walk down its corridors and ride its underground train, we seem to flow. This natural ease of motion is one reason the airport is consistently named among (and usually the) most efficient in the world by the Air Transport Research Society. Architects, engineers, and passengers can easily see that the Atlanta airport is a masterpiece of design. The more interesting question is: Why does it work?

The traditional approach to finding the answer sends us back to the drawing board. By poring over the airport's blueprints, we can determine the placement of each corridor, escalator, elevator, and concourse train. By examining the relationship between these spaces and the travelers rushing across them, and then performing thousands of other calculations, we can understand how all the parts of this giant jigsaw puzzle fit together so neatly.

After doing all that, we could build a model to help us construct our own efficient airport. This is how much of science works. Researchers and engineers catalog the elements in play in any given situation, whether it's the many components that comprise a modern airport or the complex interplay of solar heating with oceanic and atmospheric currents that contributes to global climate. Like master chefs, they continuously tweak their recipes, adding and subtracting features to their models, assigning them different weights, to come up with a model that "works."

This ad hoc approach is laborious but effective, enabling us to fathom a wide range of phenomena. It is also limited because it reflects the absence of principle. We must slice and dice the variables because we do not know the law governing their interaction and selection. It is akin to finding that our cake tastes just right when we add one cup of sugar, without truly understanding why half a cup or two cups are not the right amounts. It is like seeing the tree-shaped pattern of lightning bolts and river basins and not knowing why they look the same.

The constructal law offers a better approach. Instead of forcing us to rely almost exclusively on rejiggering variables, it is a principle we can use to build better things.

For example, the relationship between mass and speed we discussed in chapter 3, and the discovery that biological creatures should evolve to cover longer distances per unit of useful energy consumed, can help us create better cars, ships, and airplanes. The fact that the design of trees and other vegetation should facilitate the flows of water and stresses described in chapter 5 can inform our construction of mechanical structures (from beams and bridges to self-healing metals) that are also vascularized. And the insights about hierarchy detailed in chapter 6 help us understand why some social systems perish and others thrive.

These findings do not provide templates from the natural world that we can copy in our own designs. Instead, they illuminate the principle that already governs that work. Just as the discovery of the laws of motion allowed us to build better flying objects, the constructal law will jump-start our own efforts today. In this chapter we will explore how we can design better airports, roads, and cities by taking an even more dynamic look at hierarchy and the evolution of technology.

Until now we have focused on size, including the fact that all vascular flow systems generate multiscale channels because this is a good design for spreading a current from a point to an area or an area to a point. But that raises several questions: Where should those channels be placed? And in what combination? Is there a principle we can use to predict not just the fact of hierarchy but its design?

The answers emerge when we remember that the constructal law concerns movement, access, and speed. Flow designs—from lightning bolts and trees to scientific laws—emerge and evolve to facilitate flowing currents. Although these designs are what grab the eye and command attention, they are not the main attraction. Like trains and planes, they are a means to an end. They are global engines that have arisen for one reason: to enable the currents that flow through them to move more easily across the landscape.

In the hierarchy of sizes and numbers, the size of the main channel (the Mississippi, the aorta, the president) is less important than the fact that it moves the most current quicker and over a longer distance. The smaller channels of varying sizes and the interstices move less current, less rapidly, over shorter distances. As we will see in this chapter, the emergence of multiscale design hinges on the balancing of these two flow regimes. The key design principle is this: The time to move fast and long should be roughly equal to the time to move slow and short. When this occurs, currents flow with ease over the area inhabited by the entire flow structure. This is the foundation of all constructal designs, including the Atlanta airport.

Using this principle, we no longer have to rely on ad hoc models to determine the foundations of good design. Suddenly, we see the connection between seemingly unrelated phenomena, between the design of snowflakes and river basins, of flowing lava and bacterial colonies. It predicts how dogs should run along the beach; the path of your morning commute; as well as the development of Rome, Paris, and other cities.

All these connections begin with the prediction that given freedom, entities on the move should generate and seek paths that allow them to move faster and farther per unit of useful energy. Because flow is governed by this principle, this is true for the simplest forms of movement and the most complex.

In addition—and this is key—because the constructal law summarizes an evolutionary tendency, simple and complex designs are not discrete phenomena. They are part of a continuum in which smaller structures should morph *inside and along with* larger ones. This means two things. First, the complex designs are rooted in the simplest—the intricate air transport routes followed by all the planes flowing over the globe (Figure 42) grow out of the basic movement of people walking from here to there. Second, even as each component of the flow system evolves to flow more easily, it is also part of a larger system whose shape and structure are also evolving to strike the right balance among all its components to enhance its flow. To put this in human terms, we could say that the constructal law finds the nexus between individual self-interest and collective action.

To see how, start with a drawing of a straight line. It is the most direct path from one point to another when only one type of movement is involved. Unobstructed light, for example, follows a straight line between two points. Given freedom, so do people. If a man walks from *A* to *B*, and if the surface under his feet is paved uniformly, as in a parking lot, then the urge to have access inspires him to follow the straight line *AB*.

The broken line is more complicated and an even more prevalent drawing in the nature of flowing things. We see it whenever the straight-line path is unavailable, which is often the case for wingless people who cannot zip across the clear sky. Nevertheless, we seek special paths and we find them, zigging around a

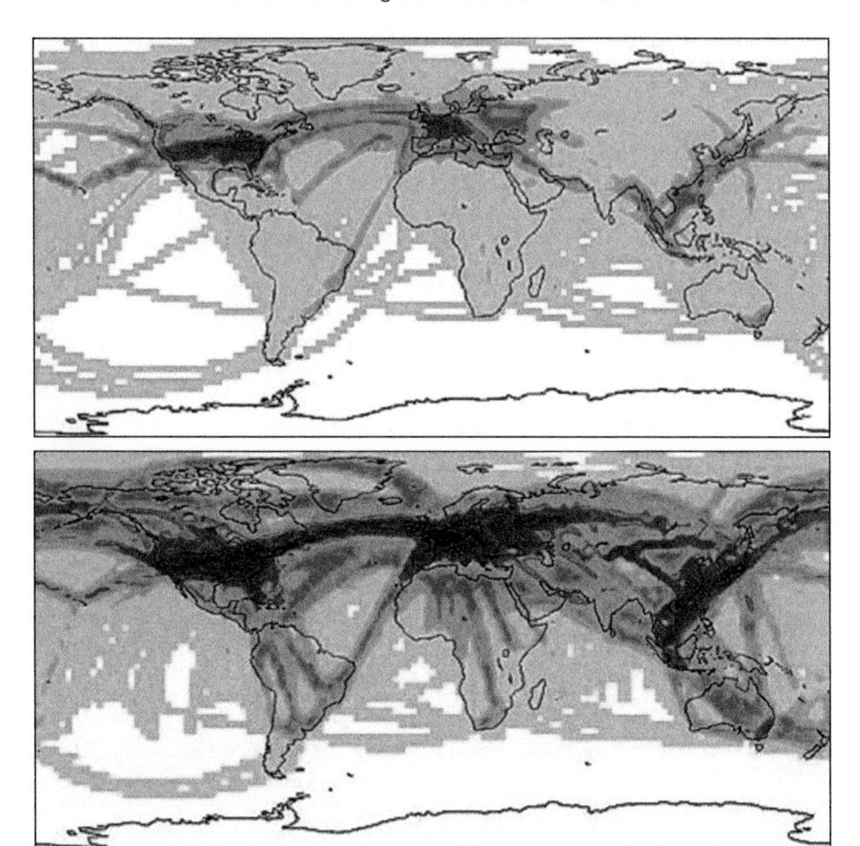

Figure 42. Where aircraft flew in 2002 (top) and where aircraft will fly in 2050 (bottom). The figure shows the density and paths of all aircraft, which are visible because of the trails of condensation left behind every aircraft.

tree here, zagging to the shortest path around a body of water there, to get where we are going.

The broken line is also the way to go when the flow between points *A* and *B* involves two kinds of movement, such as running and swimming. Tim Pennings, professor of mathematics at Hope College in Michigan, demonstrated this through a clever experiment involving his Welsh corgi named Elvis. While playing ball on a beach at Lake Michigan, Pennings noticed that when Elvis started running from point *A* on the beach to fetch a stick floating in the water at point *B*, he always chose a special point on the water's edge, point *J*, to jump into the water and swim to *B* (Figure 43). The jumping point was such that the segments *AJ* and *JB* were almost, but not exactly, perpendicular. Why? Because it is easier for Elvis to run than to swim.

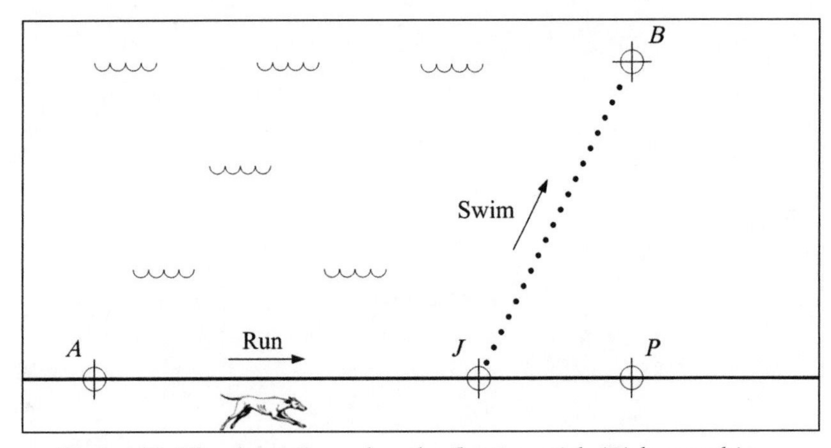

Figure 43. The dog (*A*) reaches the floating stick (*B*) by combining two very different kinds of movement, running and swimming. The dog runs on the beach from *A* to *J*, then swims from *J* to *B*. The refracted path *AJB* can be predicted by minimizing the total travel time from *A* to *B*. The same refracted path is predicted by minimizing the work done by the dog from *A* to *B*.

This does not mean that the shortest swim is the answer: That would have been the case if Elvis always jumped into the water at point *P*, which he did not. The run and the swim should be balanced against each other such that the entire effort of getting from *A* to *B* is reduced. In this way, Elvis solved one of the basic problems in constructal theory: finding the quickest way to get from point *A* to point *B* when two forms of movement are involved. The answer he intuitively discerned was that the best route is not always the shortest. Dogs of innumerable generations before Elvis learned this the hard way by catching the food floating at *B*, starving, or drowning. The same instinct is in us. If we start running in order to save a person drowning in the ocean, each of us will run only partially along the edge of the beach before jumping in, for example, from *A* to *J* in Figure 43, not from *A* to *P*.

The mind is a wonderful thing because it can imagine many variants of the movie that we just watched. Instead of a stick floating on the water, we see an imperiled person caught in a riptide. A woman leaps into action. Not just any woman, but the world-record holder in the 100-meter dash. Problem is, she is a terrible swimmer. She will follow instinctively a broken line that maximizes her running time and minimizes that spent in the water. Similarly, if our would-be hero is a champion swimmer who runs like a tortoise, she will jump into the water sooner because this is a better route for her.

The constructal law does not command a one-size-fits-all design. It proclaims that everything should seek greater and greater access in an environment booby-trapped with constraints, obstacles, and surprises. These countless constraints—such as the relative running and swimming skills of people, the topography encountered by ground water—account for much of the diversity we find in nature without violating the principle. This means that we can use the constructal law and simple geometry to predict what every path should look like to reduce global effort.

We do this by recognizing that all lines involving forward motion form angles between 90 and 180 degrees; anything less than 90 degrees is moving backward. The straight-line path, for example, is a 180-degree angle. When this path is not available, the line is bent or broken. How much? The answer depends on the ratio between the speeds of the two types of movement involved—V_0 (slow) and V_1 (fast). The greater the discrepancy, the smaller the angle. To determine the broken line that Elvis should follow, we divide his running speed by his swimming speed. If they are identical, he should follow a straight-line path to the stick. As they diverge, his angle of "refraction" should decrease. The extreme design—the 90-degree angle of refraction—is the natural way to flow when the two flowing modes differ significantly. We see this most clearly in modern cities, where there is a vast difference in the speeds of the two most common forms of movement, walking and riding in cars. Thus, the avenues and side streets meet at 90-degree angles.

Elvis the dog had not crossed my radar in the early 1990s when I was designing advanced cooling systems for electronics. But it turns out that he had found a solution to a simple version of the far more complex and prevalent challenge that confronted me. With all due respect to Elvis, he had to worry only about himself, how he should combine running and swimming to fetch his stick. My work addressed the far more common broken-line problem in nature of how to minimize the travel time involving two flow regimes and innumerable currents over an *entire area.*

Put another way, my work did not just concern Elvis but all the dogs and all the sticks on the same area. Or, more precisely, how to ferry all the heat generated by all the electronic circuits out of a tiny, confined space—just as the Atlanta airport must accommodate all the people traveling across its concourse, and the air transport system must facilitate the flow of all the planes around the world. As we saw in chapter 2, I accomplished this by placing the circuits on material that did not conduct heat well (a slow way of moving akin to swimming) and then placed a strip of highly conductive material down the middle (a faster way of moving, like running).

The heat generated by the circuits spread diffusely—slowly in a disorganized, patternless movement—over the short distance perpendicular to the center strip, where it moved relatively quickly over the long distance to the heat sink. I found that I could cool a larger area by adding more strips of high-conductivity material in a treelike pattern. How many trees? My answer provided the breakthrough that illuminated a central principle of good design: Just as Elvis balanced the time he spent swimming and running, the right answer was one that balanced the resistances that all the heat encountered while moving in these two flow regimes.

The fact that Elvis and I derived the same answer to a similar problem offers further proof that a principle is at work. As we have seen, when a straight-line path is available, nature embraces that. It is only when that ideal path is not available that we find a broken line—not just any broken line, but a special one that still bends toward the fastest route. When that broken line involves two types of movement, its shape will strike a balance between them in a configuration that provides better access for whatever is flowing. These three scenarios generate simple drawings because they concern the movement of a single current between two points. Flow in nature, however, is almost always more complex because it usually involves vast quantities of mass—all the water in a river basin, all the elephant mass over land, all the goods and services produced by a local or global economy. Thus, the design generated is not a single straight or broken line but a system of multiscale channels with hierarchy, that is, a superposition of broken lines. It, too, is governed by the constructal law.

So far we have focused on the evolving design of the channels that carry that current. But the constructal law governs all flows over an area or throughout a volume. Think of a river basin. The eye is drawn to the vascular structure of its channels. But that is only part of the drawing. Equally important is the area between those branches that feeds water to them. Just as airports do not generate passengers but channel them from surrounding areas, and banks do not generate money but collect it from thousands of depositors, the river basin must collect its water from the surrounding ground. Similarly, my strips of high-conductivity material did not generate heat but attracted it from the surrounding area or volume.

In terms of a drawing, then, the entire area of a flow system is both the black (the lines of channels that emerge) and the white (the rest of the area—the interstices—that serves or is served by those black lines). When we take a global view, we see that everything that moves over an entire area has two ways to flow—short and slow (the white), or long and fast (the black).

The constructal law predicts that currents should move slow and short when that is the better way to flow, and fast and long when this improves movement. That is, in every instance the flowing current should "select" the better mode for flowing. As we saw in chapter 2, in many cases this involves the transition from laminar to turbulent flow in water, air, or other fluid. In larger designs, this means that current will flow diffusely until it can move more easily by coalescing into channels, and vice versa. Water seeps through the ground until it encounters more water; the water then can move more easily by forming rivulets and larger streams.

As these hierarchical designs evolve, generating multiscale channels (few large, many small) to cover more ground, they should strike a balance between these two flow regimes every step of the way.

The river basin does not begin with the Mississippi but with water seeping diffusely in the ground that forms the first rivulets; the evolution of our circulatory system did not begin with the aorta but with the diffuse flow of blood in tiny organisms that eventually generated capillaries and larger blood vessels; our transportation systems did not begin with highways and airports but with the flow of people across untamed ground that created the first footpaths.

The water in the river basin combines the slow seepage down the hill with the fast flow along the river channel. The slow flow is perpendicular to the fast, just as the access lanes from the concourses in the Atlanta airport are perpendicular to the trains they lead to. Scientists who study the movement of chemically marked water flowing down a river basin are finding that the seepage time down the hill is essentially the same as the time spent in all the river channels. This occurs in all the river basins on Earth.

As we detailed in chapter 1, it also happens in the design of animal lungs. Take a breath, a nice deep one. The time needed by the air to flow quickly down the airways from your mouth to the alveoli is the same time required by the oxygen to diffuse slowly across each alveolus and into the tissue where it is absorbed by the blood. Notice, too, that this rhythm, this design—these two times in balance—remains in place no matter how quickly you are breathing; the time spent inhaling is always the same as the time spent exhaling. The same rhythmic design characterizes the beating heart, blood circulation, digestion, excretion, ejaculation, etc., and is predictable based on the constructal law.

In fact, this design happens everywhere. The surface of the Earth is a tissue woven of all these flowing things. It is covered by an extremely diverse number of things that flow and move in these two different ways. There are, of course, many poor designs, especially in the man-made world, whose evolutionary history is much shorter than that which we find in older, more entrenched systems. More telling is the fact that so many of them strike the same balance between the two flow regimes we find in other natural designs.

Now we will see how, by looking at an array of engineered designs in history. We begin with one of the oldest and most famous structures, the Pyramids in Egypt, and other ancient sites that still intrigue us with their size and geometric form. Even by today's standards, their size is immense and their form is perfect. These designs are so impressive that our culture tends to attribute them to an ancient scientific base of knowledge that was lost, and to presumed links between ancient peoples living on opposite sides of the globe.

The constructal law solves the mystery of the Pyramids in a surprisingly direct way: They are the result of a universal natural phenomenon that governs the movement of all materials on Earth. This view does not take anything away from the achievements of the ancient builders. Rather, like the evolution of the wheel described in chapter 4, it is a physics argument that what our ancestors chose to do is *natural*, that to engineer is natural, to tend to migrate on the globe is natural, and that the geometry of all material flows (animate and inanimate) can be reasoned based on a single principle.

In the making of a Pyramid, the constructal law calls for the expenditure of less work by striking a balance between the time to move slow and short and the time to move fast and long. This principle accounts for the *location* and *shape* of the edifice. First, the location is in the middle of the quarry, because less work means a shorter sliding distance between the place where stones are mined and the construction site. (As technology has evolved, work sites have moved farther from the sources of material because it takes less effort to transport them.)

This same phenomenon also accounts for the shape of the Pyramid. The French architect Pierre Crozat demonstrated that builders used the Pyramid slope to move the stones upward using wood levers and ropes. Each stone was lifted, moved horizontally, and then dropped at the next higher level. In a pile of stones held together by gravity (dry-stone construction), shape means the base angle. A good angle is one in which the work spent on moving the stone horizontally is roughly the same as the work spent on moving it on the incline.

If the flow of stones is configured such that the edifice is constructed with less and less expenditure of useful energy, then the shape of the pyramid (the angle at the base) is unique, size independent, and dictated by the technology of the era. The prediction is that the pyramid construction must proceed layer by layer such that the pyramid is geometrically similar to itself during its growth (in layers, like an onion).

Put another way, there are two "media" through which the streams of stones flow—two mechanisms—one with low resistivity (moving the stone horizontally, which is relatively easy) and the other with high resistivity (moving on the incline, which is much harder). When the two media are highly dissimilar, the angle at the base (that is, the angle of refraction of the ray of moving stones) approaches 90 degrees. Rivers, stones, and animals flow with configurations that come from the same principle. We should also note that the law of refraction governs the movement of goods in economics, where it is known as the law of parsimony. For example, to ship Lucky Strike cigarettes from Durham,

North Carolina, to soldiers in Dunkirk is not to send them on the shortest (geodesic) line that links the two cities. It is to send the goods along a less expensive path if possible, which could be the refracted ray consisting of the short and high unit price (land route, by truck from Durham to Savannah), followed by the long and low unit price (by ship, from Savannah to Dunkirk).

The history of the development of trade routes documents this constructal design tendency. We often hear that a city or harbor grew because "it found itself" at the crossroads—at the intersection of trade routes. In fact, it works the other way around; the efficient refracted routes defined their intersection, the city, the port, the loading and unloading site, etc. More complicated flows are bundles of paths, refracted such that local and global flow is enhanced. A river basin under falling rain is like an area inhabited by people: Every point of the area must have greater and greater access to a common point on the perimeter. There are two media, one with low resistivity (channel flow; vehicles on the street) and the other with high resistivity (seepage through wet riverbanks; walking). The shape comes from the tendency to facilitate flow access.

The Atlanta airport is a more complex design than the Pyramids. Instead of just balancing the movement of stones, it must accommodate all the passengers and their goods (Figure 44). Movement in the airport is between a point and an area—for example, from the airport entrance to all the gates, or from one of the arrival gates to all the other gates and the exit. To cover its entire area, passengers must combine two movements. They travel slowly over a short distance as they walk down the concourses. Then they proceed quickly over longer distances by riding the train that links the concourses.

With these parameters in mind, Lorente and I used the constructal law to predict what shape the Atlanta airport—and all such designs—should have to provide the most access to the currents that flow through them. Through this work, we developed a formula that can be used to determine a good design for areas covered by two flow regimes, and thereby reduce our reliance on models.

The airport area can be shaped so that it facilitates access for every body and every thing that moves. We reasoned that in the sketch shown in Figure 44, the rectangular area ($H \times L$) is fixed, but the shape of the rectangle (the ratio H/L) may change in the minds of the designers who are seeking to facilitate flow access.

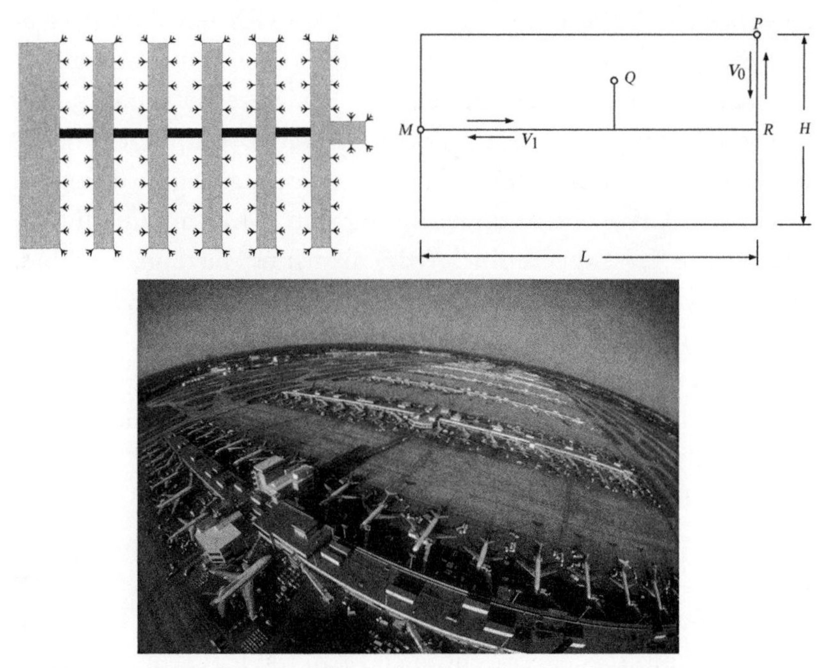

Figure 44. Two ways to flow are better than one. A large airport without trains, or without walking, cannot compete on the same area with the design that combines walking with riding in a vehicle. The Atlanta airport is a modern illustration of the seed from which all forms of urban and natural flow networks have grown. On a fixed area ($A = HL$) with variable shape (H/L) and two speeds (walking V_0 and train V_1), the time of travel from P to M (or from all points Q to M, averaged over A) is minimum when the shape is $H/L = 2V_0/V_1$. The walking time (PR) is equal to the riding time (RM). The long and fast travel is balanced with the short and slow travel.

These minds are numerous, not one. They visualize all sorts of flows—people, luggage, food, waste, services, etc.

But for now we will just consider passengers. What is the best ratio for H/L? Consider the travel between the farthest gate (P) and the terminal (M). Think of P as the least advantageous position because it is the farthest. The passenger from P must combine some walking at speed V_0, on the short side, with some riding on the train, at speed V_1. This passenger needs the time $t_0 = (H/2)/V_0$ to walk, and the time $t_1 = L/V_1$ to ride. The total time needed by this passenger is $t_0 + t_1$, and it is minimal when the shape of the area is $H/L = 2V_0/V_1$. The ratio of V_0/V_1 is the walking speed divided by train speed and is considerably smaller than 1. Consequently, the aspect ratio H/L must be smaller than 1 (as shown in Figure 44).

It is especially telling that we also discover this special shape if we take all the passengers into account. If we calculate the time to walk and ride for the arbitrary passenger Q, and if we average this total time over all the passengers (that is, over the rectangular area), we find also that the averaged time is minimal when $H/L = 2V_0/V_1$. The aspect ratio of the rectangle is a number comparable with 1 but smaller than 1, for example 1/2, because the V_0/V_1 ratio for the walker and the train is a number of order 1/4. This shape is evident in the actual layout of the Atlanta airport.

The coincidence that the best airport shape for the farthest traveler (P) is the same as the best shape for a community of travelers as a whole is worth thinking about. It raises the question of whether the airport designers behaved *altruistically* by shaping the airport to help the passengers who must use the most peripheral gates, or the same designers as a group behaved *egotistically* by imagining themselves in that airport, at every possible gate position Q. The more plausible interpretation of the final design is the *egotistical* route. Configurations emerge naturally in areas and volumes where there are large numbers of moving individuals and each individual has the same tendency, the same drive, as his or her neighbors: to seek and find access, and go with the flow. The urge to organize is selfish.

Even more amazing is that when the airport shape is the one that everybody likes, then the average time needed between the area and the point (M) is divided roughly equally between the time spent walking and the time spent riding on the train. The genius of the airport's design is this: The time to walk along half a concourse is about the same as the time to ride quickly on the train, end to end—about five minutes. That is, the time we move slowly (walking) is equal to the time we move fast (riding the train). Passengers do this through a design that should be familiar by now: It is shaped like a tree.

It is no surprise, then, that the newest design found in the world's leading airports—the most recent evolution of the flow systems we call airports—look more and more like the Atlanta airport. The facilities in Singapore, South Korea (Incheon), Hong Kong, and Tokyo (Narita Terminal 2) are all characterized by the right combination of pedestrian concourses and perpendicular trains. Through the constructal law, the evolution of airport architecture and technology is predictable. This is not copying the Atlanta airport. This is natural evolution, in accord with the constructal law.

We see a similar balance in the transport systems connected to airports. The time to fly along one of the air routes of Europe is comparable with the time to travel on land, perpendicular to the route (Figure 45). It takes roughly two hours to fly from Paris to Madrid and about two hours to drive from the Madrid airport to a locality on the entire area served by the airport. A fast train in Europe links two neighboring cities within one hour or less. This is also the time needed to travel between home and the more numerous train stations. Inside the city the areas covered are smaller but the principle still reigns. The minutes needed to get to the train station should be comparable to the minutes spent on the train. In other words, you would be less likely to drive

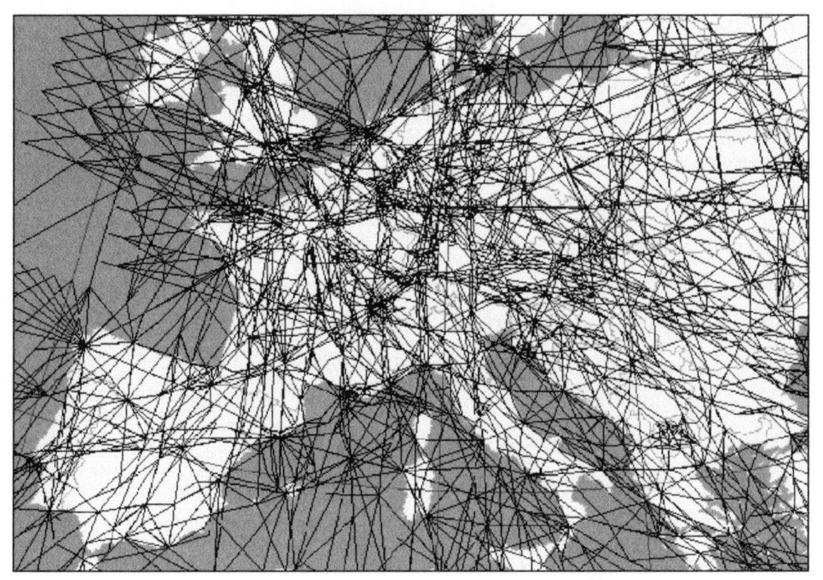

Figure 45. Tapestry of air mass transit over Europe. The burning of jet fuel is for moving people and goods on the whole area: This flow is hierarchical and nonuniformly distributed. Large centers and thick channels are allocated to numerous smaller channels. The fine channels are allocated to area elements (between the channels) that are covered by ground movement—people, and all the animate and the inanimate flows of the environment. The time to travel long and fast (along the channels) is comparable with the time to travel short and slow (across the areas between channels).

two hours to a train station to take a half-hour train trip. Even if it saves you a little time in the long run, you would probably just continue driving the full distance to avoid expending the effort to find a parking space, buy a ticket, and overcome other forms of resistance. In a well-designed transportation system, we spend equal times on our feet and on our bottoms. Just as in the Atlanta airport. In a poorly designed system, these times are out of balance.

As the TV ad says: Individual results will vary, depending on the person's starting location. The principle applies to the average time it takes *everyone* to complete his or her journey; it describes the flow design of huge numbers of people, not the experience of each person. The constructal law is the big picture, but it is also the small one. It is the forest and the tree. It is all the animal mass that moves on Earth and the athlete that runs quickly.

The shaping of the airport area is a template for the shaping of all the other slow and fast loops in the flow tapestry of nature. For instance, we can use it to predict the design of cities—which serve many functions but whose shapes and structures are determined by the need to enable people and goods to move easily. Anywhere, and everywhere, it's all about flow.

The smallest street with its houses, lawns, and yards on both sides is just like the airport area ($H \times L$). This is the smallest building block of the city design. The shape of the smallest city block is dictated by the ratio $2V_0/V_1$, where V_0 is fixed (the walking across the lawn), and V_1 increases in time because vehicles become faster in time.

The entire fabric of city design evolves in time because of *technology evolution*, which has reduced the time it takes to cover an area. With pure theory in our minds, we can look back at the evolution of city designs and marvel at how continuous the running of this movie has been. In antiquity, the speed of the heavy cart pulled by the ox was about twice the speed of the human. This means that the smallest street would have emerged naturally on a square city block—an area with a shape H/L comparable with 1. Given the fact that the simplest drawing for the individual house and yard is a nearly square rectangle, this means that in antiquity the smallest city block must have had one or two but not many more houses on one side of the street. Today this holds true for all the other rural and urban areas where the fast mode on the tiny street is the ox and cart.

Fast-forwarding to the present, in the design of cities, where the car speed is more than ten times the walking speed, we should expect more houses on the smallest streets. In addition, city blocks should become more elongated as the technology of transportation evolves. This is in agreement with designs of modern urban developments.

The two ends of this movie are illustrated beautifully in one map of modern Rome (Figure 46), the birthplace of Western civilization (city living, literally). The center of Rome is the ancient city, and here the streets are considerably shorter than in the more recently built, peripheral areas (for example, the upper corners). History begins to make sense now that the principle of the evolution of design is known.

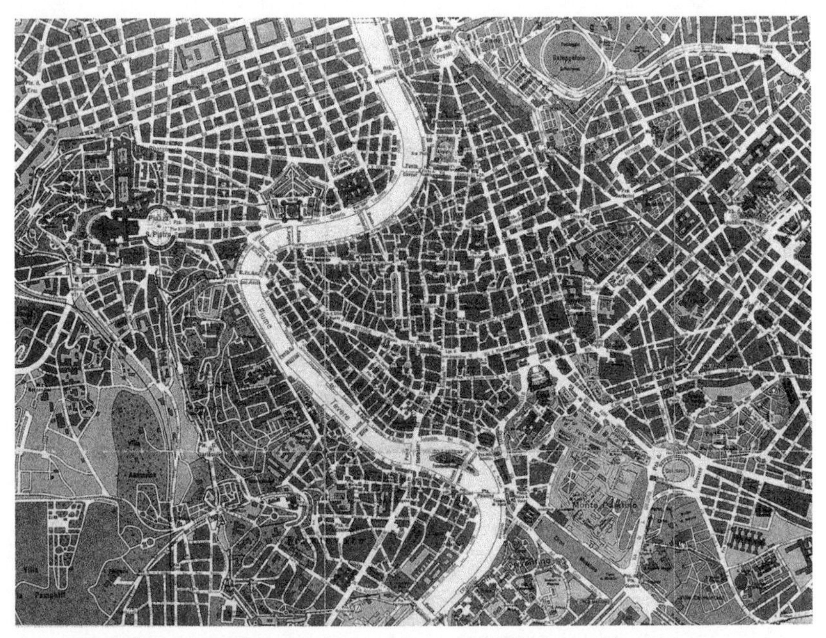

Figure 46. The plan of modern Rome, showing that in the ancient city (the center) the street length scales are considerably shorter than in the new outskirts.

All roads lead to Rome. This is how the population moving from the countryside (the area) was connected to Rome (the point). Not a radial pattern with roads in all directions, but a tree-shaped one with only a few major arteries leaving the city. This natural design connects every large and small city to its allocated area, and all city and area units are connected to Rome. The same design connects every river basin and delta to their points of discharge and supply.

As the human settlement becomes larger in time, streets and patterns of streets emerge. Small villages have only two or three streets that touch the main square and branch outward into more roads that cross the land. Larger towns and cities have grids. The reason is that in larger settlements there are many more points of interest that must be connected by tree-shaped flows of humanity to the surrounding area.

The street design is a plaid: A few broad streets form a grid that is superimposed on a grid formed by narrower streets. To see this, imagine that inside the city area the entire population (all the Qs in Figure 44) must have access to one destination, M (for example, a church). The design that serves the population is a combination of slow and fast movement (small and large streets), as in the Atlanta airport. The movement of this population between the area, $H \times L$, and the point, M, is tree shaped.

Next, imagine another destination point, perhaps a market, in the same area. The area-to-point flow must be another tree-shaped flow, but this new tree will be superimposed on the preceding tree flow. More points of flow attraction on the same area will require more tree flows superimposed on the preceding ones. The infrastructure of *solid* channels that facilitates all these possible (superimposed) tree flows of humanity—whenever they may occur—is the grid of streets.

All the individual area-to-point movements through the grid, however, are tree shaped, not grid shaped. But, through the superposition of many area-to-point movements, it begins to look like a grid. Imagine the movement of the city population to a political rally in the main square. Citizens converge in river-basin fashion, the many from the small streets becoming groups and columns on the avenues. Another tree-shaped flow is the morning migration of commuters from their homes to the train station. In the network of air routes over Europe (Figure 45), the flow of passengers from (or to) a city is tree shaped, not grid shaped. How do the streets and the air routes accommodate all these superimposed tree-shaped flows? By evolving into a superposition of trees, which is a grid, that is, a network.

Lovers of truth in language will note that we are not calling the airport and river-basin flows "networks." A tree is not a net. One does not catch fish with a fork or a broom. A grid is a network because it has loops, as if it were woven into a net. The grid is a net because it is a superposition of actual flows, which are shaped like trees.

As the city evolves, its population increases and the street pattern evolves to serve the growing population. The grid is a sign of the evolution of civilization: living inside a walled area with many objectives distributed as discrete points on that area (markets, churches, schools, government buildings, train stations, etc.). The grid of streets is the architectural invention of Hippodamus of Miletus, who designed the city of Rhodes in 408 BCE.

City designs continue to evolve as populations increase and transportation technology improves. Because highways and automobiles have become much faster and more economical (per kilogram transported) than their predecessors in the era when the city center was built, the city population finds itself in agreement on adding two modern features to the design. Suspended highways and underground tunnels are built across the city, passing through the center. Circular highways are built around the slow-moving center, famous examples being Le Boulevard Périphérique around Paris, and the Beltway around Washington, D.C.

Theory empowers us to expect these signs of evolution in the future. As the large city expands significantly beyond its beltway, and as highway and automobile technologies improve, a second beltway (wider, faster, with a radius twice the original radius) will emerge around the city and the first beltway.

Taking a step back, we see that these designs have emerged naturally as humans have been maintaining, on the whole, a balance between movements that are slow and short and fast and long, every step of the way. When we recognize and use this principle, we can fast-forward our efforts to design better transportation systems. To see how, consider a broad sketch of all the people who travel only on local roads to get to work. For them, the time they spend walking from their house to the curb (1) and driving out of their cul-de-sac (2) is their short and slow movement. Their time speeding down the avenue (3) is their fast and long.

For all the people who must also travel on the highway (4), those first three movements are their short and slow movement while their time barreling down the interstate is their fast and long. Similarly, for all the travelers taking to the skies, those first four movements become the slow and short while their time on the plane is the fast and long. As urban planners design new plans for multimodal forms of transportation—networks of interconnected sidewalks, bike paths, roads, trains, ferries, airports, etc.—they will build better systems by keeping all these modes in balance.

Knowing the principle also warns the designer about what not to do. On the East Coast of the United States, there is much talk from proponents of high-speed trains like the TGV in France, and the light-rail trains like the RER in Paris. These ideas sound good, but on the design of the American landscape they make as much sense as the military saying "hurry up and wait." Why use a bullet train when at the other end you have to wait one hour for a bus, and where there are no sidewalks and safety at night? This underscores the folly of trying to impose a design that has not evolved naturally. Building a high-speed train where there is little supporting infrastructure makes as much sense as placing a larger river near a hillside without streams to feed it.

There must be a balance between the time spent traveling long and fast and the time of moving short and slow. When this balance is reached, large numbers of citizens (future users) vote for the design. They vote in the booth, telling the city to build it. They vote with their wallets when they pay city taxes and buy tickets. This is why the flight from Washington, D.C., to Raleigh-Durham takes half an hour—the same time that most of us need to drive home from the airport.

TGV, yes, but why?

The balancing of times, and the search for greater flow access for everybody, are mental viewings that take the designer well beyond the two-dimensional (on an area) examples discussed in this chapter. The same ideas work in three dimensions. A tall building works well when its elevators are fast enough so that the time spent on the vertical is comparable with the time walking in the corridors. Security checkpoints in the airport and in war zones work best when the time and effort spent in the bottlenecks are comparable to what is spent en route to them. The infrastructure and security of a previously virgin or newly liberated area owe their design to the same principle as all other design phenomena.

The designs of the Atlanta airport, Rhodes, and Rome were not copied from nature. Their emergence and persistence as living flow systems are nature itself. Now we know the principle that underlies their repeated occurrences, scaling rules, and longevity.

Who designed and who built the patterns is not the question. Science is not the search for a designer. Huge numbers worked and continue to work on the design, and they use time and freedom to make changes, and memory to construct it. Culture serves as memory in the evolution of urban design. The dry riverbed and the seismic fault are memory for river-basin evolution. The new scientific aspect that unites all flow systems is that they possess design (pattern, configuration, shape, structure), and the design-generation phenomenon is universal and anticipated by the constructal law.

Now we see why it is useful to know the principle. Designers are empowered by it. Their imagination leaps ahead, over the territory that would have been littered with tried and rejected designs. The traditional first move on a designer's table is to look and to copy. Looking at nature and copying what millions of flow systems have built is called "biomimetics." It works only when the person who looks understands the phenomenon that generated the natural drawing.

Thus we render biomimetics obsolete because the constructal law allows us to predict and explain the designs that emerge naturally. Looking at the drawings in a handbook is the most common approach, and it leads to marching in place, not to leaping forward. Copying an inventor's revolutionary design is much more effective, but such leaps are either costly or illegal. With the constructal law we are the inventors.

CHAPTER 8

The Design of Academia

Evolution has long been an idea in search of a principle. A concept as old as science itself—Aristotle, for example, suggested that nature was ruled by a desire to move from lower to higher forms—"evolution" has been invoked through the millennia to describe change over time. Nowadays, this single word encapsulates Darwin's work about biological life and the subsequent research that has refined and elaborated his insights. It is also employed much more loosely to describe the development of just about everything. Library shelves sag from the weight of tomes describing the "evolution" of science, nations, written languages, and social values; of religion, war, technology, art, cooking, and even the beautiful game of soccer.

This history is the story of good hunches. Every discernible thing, every design in nature, does evolve. It is dynamic, not static. What has been missing is the single principle of physics that unites these phenomena and allows us to predict how they should evolve in the future. Using the constructal law, we recognize that not only biological species but also technology and language, religion, education, and all the rest are flow systems that configure and reconfigure themselves so that the bodies that possess these designs (we, the cultured) move more easily on the globe. It shows us that evolution is far broader than Darwinians have believed and far more specific and powerful than other thinkers have imagined.

"Evolution" means design modifications over time. How these changes are happening are *mechanisms*, which should not be confused with the principle, the constructal law. In the evolution of biological design, the mechanism is mutations, biological selection, and survival. In geophysical design, the mechanism is soil erosion, rock dynamics, water-vegetation interaction, and wind drag. In sports evolution, the mechanism is training, selection, rewards, and the changes in the rules of sports competitions. In technology evolution, the mechanism is innovation, technology transfer, copying, theft, and education.

What flows through a design that evolves is not nearly as special in physics as *how* the flow system acquires and improves its configuration in time. The *how* is the physics principle—the constructal law. The *what* are the currents and the mechanisms, and they are as diverse as the flow systems themselves. The *what* are many and the *how* is one. Hierarchy more simple than this does not exist.

The constructal law advances our understanding of evolution by proclaiming that design should emerge across nature to facilitate flow. It also holds that these configurations should morph with a clear direction in time: to provide better and better flow access. Evolution, then, is *measurable* in terms of how much easier and farther things move on Earth.

As it predicts why design should emerge and evolve, the constructal law reveals the broad patterns that abound in nature. Despite their great diversity, flow systems faced with similar challenges and constraints tend to acquire similar designs. Inanimate and animate designs evolve as if they are "intelligent," because they appear to come up with the same answer to the problem of how to flow more easily. They also generate the same designs that we come up with to facilitate flow; that is why their designs are predictable.

Pattern generation is evident in the predictable design that emerges among the various components of a flow system. The vascular, hierarchical designs we find throughout nature strike a balance between the speeds of their currents (each of which selects the mode of flow, slow and fast, that works best for them) by generating multiscale channels. It is also apparent in the evolving design of even larger flow distribution networks, including the hierarchical distribution of tree sizes in the forests and in the emergence of human settlements—a few large cities, with many small communities on the map.

This overriding natural phenomenon has been noted by researchers. But they have described it empirically, as power-law correlations, hierarchies, allometric scaling, and Zipf distributions of frequency versus rank. Researchers have observed but could not predict. They have known the *what* but not the *how.*

This *how,* the constructal law, sparks especially surprising insights when applied to social dynamics. The prevailing view holds that the institutions built by humanity are subject to the desires of people, not the laws of nature. This view is wrong, not just philosophically and not just as a practical matter. It is wrong as physics, because social designs emerge and evolve as a result of the selfish urges of many individuals who do not consult one another. Each has the tendency to flow more easily; all find it is easier to flow together, with design. This means that social designs, like other flow designs, occur naturally.

We have already seen how the constructal law predicts the evolving designs of engineered entities such as the wheel, roads, airports, and human settlements. In this chapter we will focus on two areas that seem less concrete: academia and human relations. We will see why they, like every other social system, are hierarchical designs that evolve to cover an area with current, and manifest the same patterns that emerge in other natural phenomena.

Each new release of the rankings of America's best universities by *U.S. News & World Report* is the talk of the campus. Some administrators discount the importance of rankings, while the rest declare that the university is finally (now) poised to execute "the great leap forward." This reaction has not changed in years, because the rankings have not changed in any meaningful way in years.

The usual suspects—Yale, Harvard, Princeton, MIT, Stanford, and Duke—are annually reconfirmed as the leading national universities; Williams and Amherst remain among the highest-ranked liberal arts colleges. The graduate school rankings—Yale and Harvard tops in law, Stanford in business—continued this dog-bites-man story.

There is often a little more movement down the lists—a few schools rise a notch or two; others fall slightly. But for all the concern about administrators gaming the system, the absence of change stands out. To quote the Talking Heads song, with academic rankings, it's the "same as it ever was."

Why do the rankings seem carved in stone? The answer cannot be found by studying the metrics used by *U.S. News* editors but by applying the constructal law. As we have seen, a pattern that persists in time by resisting big forces to change speaks of the much bigger forces of nature. It speaks of physics and the design of nature, of the evolution of shape and structure that facilitates our movement on the globe. Yes, you read it correctly—science and education facilitate our movement. The urge to move more easily is what drives the tendency to acquire knowledge, not the other way around. Without science and education we would still move but not much, because we would be hiding in caves. Because knowledge and information are currents that enhance our own movement, they acquire evolving design in accordance with the constructal law.

It should be no surprise that the architecture we find in river basins and forests (few large, many small) is the same one we see in our system of higher education—a tiny number of top-ranked universities, a few more second-tier institutions, and many lower-ranked schools. What may seem remarkable is that this hierarchy is as rigid as the one we find in those other "natural" systems.

How come?

We answer this question with pure theory, by predicting that education is an evolving global flow system with design that is governed by the constructal law. The rankings are an expression of this phenomenon. We begin our analysis by putting aside the Darwinian interpretation of the rankings, which casts individual schools as competitors in an epic struggle for survival/supremacy. Instead, *all* the colleges and universities are components of the single larger flow system that covers the entire globe. Just as the Mississippi River is not competing with its tributaries but working hand in glove with them to move water, the schools of various rank—both high and low—together form the river basin of education that spreads knowledge across the global landscape.

Next, we identify what current is flowing through the design. The answer is ideas, and the pedigree that the density of ideas attaches to those touched by the flow. A scholar and a university become known because of the ideas they generate. Good ideas travel and persist (to "persist" means to keep on morphing and traveling from those who know to those who need to know). The good ideas are the ideas that are adopted by others worldwide. Most ideas are replaced and forgotten; like the vast majority of published research papers, they are not even noticed.

If ideas are the current flowing through the academic system, what is the measurable characteristic that makes one school more highly ranked, or "better," than another? That is, if we rank cities by the size of their populations (Paris is a bigger channel than Lyon), what do the highly ranked schools possess more of than their lower-ranked brethren? It is certainly not their physical sizes. The top schools do not have the most students. It is, instead, the visibility, the fame, the usefulness of the ideas they generate. In education, fame, or visibility, is synonymous with greater access through the vascular structure of societal flows. Students flock to high-ranking schools because they know these schools can help them enter the main channels of society. Education flows in one direction: from those who have it to those who seek it. When both ends of each such river basin have it and know it, the flow stops. What is not news does not travel.

Seen constructally, a university is not the piece of land in a particular spot. It is the professors, their disciples, and the disciples' disciples. It is the ideas that flow through these human links and into the books of our evolving science and culture with which we walk on Earth. Because hierarchy occurs at every level of the design, each university is the central node, heart, and aorta, nourishing and sustaining its students and others with the ideas it generates.

It is also a channel on the entire world map, a component of the highly complex global vascular flow network of knowledge. In time, this global vasculature evolves like a river basin during the rainy season: All the streams swell, but their hierarchy remains the same.

The historical institutions—from the Universities of Bologna and Padova to the Sorbonne, Oxford, Cambridge, Coimbra, and Harvard—have earned their rankings in this global system because of the fame of the ideas they have and continue to generate. This entrenched hierarchical design persists because it facilitates the flow of ideas across the world.

With these ideas in mind, the constructal law predicts that all the universities should generate a hierarchical design to facilitate this flow. That is, they should produce the distribution of design features (in this case, of universities) that we find in the design of river basins, forests, and other natural phenomena. The ranking of these schools should be based on the fame, the usefulness, of the ideas they generate.

To test this prediction, I took the only unbiased measure of academic visibility available in my field of engineering—the number of citations of an author's creative output compiled by the Web of Science. It is an unbiased sample because the researchers who cite an author's work do so because they read it, valued it, and used it. These numerous voters are not recommended by anybody. A magazine does not handpick them. They do not belong to a club. The best part is that one can see who they are and why they cited the author.

For each U.S. graduate engineering school ranked in the top 50 by *U.S. News*, I counted the number of names that appear on the most-cited list. I plotted this number on the ordinate in Figure 47. The abscissa indicates the ranking in *U.S. News*. This figure provides a bird's-eye view of where university rankings come from. The highly ranked engineering schools are homes to researchers who are highly visible. The lowly ranked schools are not. The left end of the scale is dominated by schools with ordinates in the 5 to 10 range. The right end is dominated by schools with 0 on the ordinate.

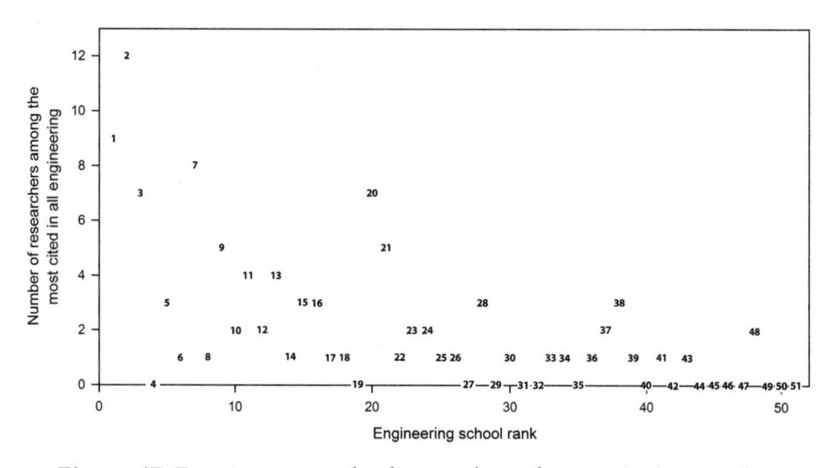

Figure 47. Fame versus rank: the number of most-cited researchers in each top U.S. engineering university versus the rank of the engineering university.

This is not a chicken-and-egg argument. The direction is one-way. The university rankings come from the highly cited, not the other way around. Authors are famous because of their creativity, not because of the name of their employer. In my own field, we cite Ludwig Prandtl all the time because of his boundary layer theory, not because of the fame of his employer, the University of Göttingen.

The scatter in Figure 47 does not diminish the firmness of this conclusion. One can argue that "size matters," which is why some highly ranked schools (those ranked numbers 4 and 19, for example) do not have any high-cited researchers. These examples are the exception, not the rule. To stress this, I replotted the points of Figure 47 by scribing the same values on the ordinate, and using a new abscissa: the rank of the particular school on the list of the most-cited researchers in all engineering. The result is Figure 48. For example, rank 1 on the abscissa of Figure 48 belongs to the school with the most names on the most-cited list (that school was ranked 2 on the abscissa in Figure 47). Because of the logarithmic ordinate in Figure 48, the points with 0 on the ordinate are not shown.

In the new representation of Figure 48, the points descend smoothly from left to right, producing the same pattern we find

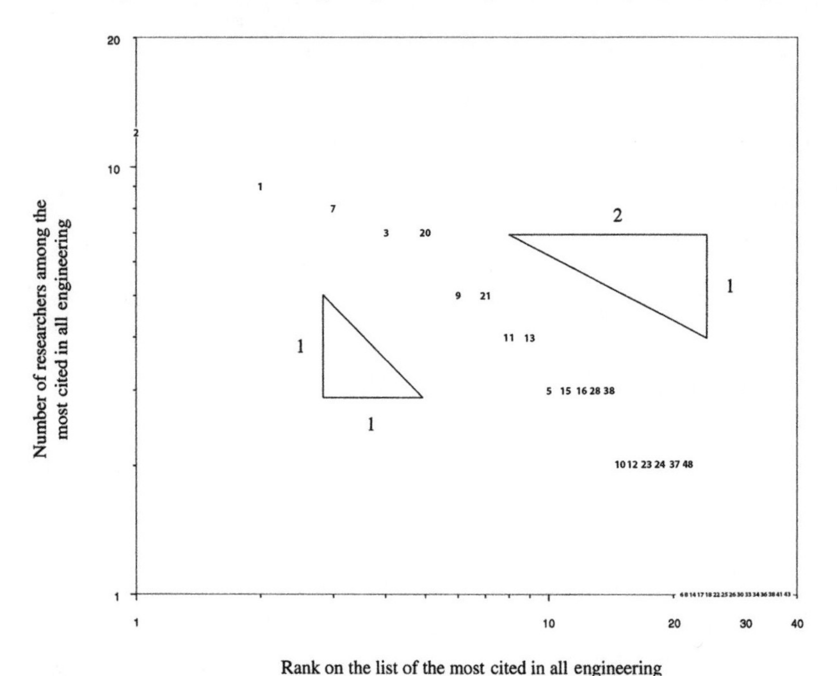

Figure 48. The number of most-cited researchers in each top U.S. engineering university versus the rank of that university on the most-cited list.

in other natural flow systems. Practically all the points that were on the left in Figure 47 are still on the left in Figure 48. Immobility also characterizes the points on the right in Figures 47 and 48. The 30-to-32 abscissa range of Figure 47 is essentially the same as the 25-to-40 range of Figure 48.

Figures 47 and 48 show that the ranking of universities is hierarchical, like the airways of the lung, the channels of the river basin, and the cities of a country or continent. The more highly ranked, the fewer the candidates for the high positions. The trachea, the Danube River, and Paris are not to be confused with the other airways, river channels, and human settlements. The opposite is true in the other direction: The lower the rank, the more numerous the potential candidates; hence we see more apparent movement the farther down we go on the *U.S. News* rankings. Why?

The clue lies in the nearly straight line that the data form on the log-log plot in Figure 48. This line has a slope between $-1/2$ and -1 and is coincidentally the same as the distributions of city sizes throughout the modern history of Europe (Figure 41). The similarity between Figures 48 and 41 suggests that the distribution of sources of knowledge is intimately tied to *geography*, *geology*, and *history* (to the evolving drawings of the flows on the landscape), and to the tissue of information channels on the surface of the globe.

This insight allows us to take another step in our constructal view of education. So far we have shown that the fame (usefulness) of the ideas generated is the current that flows through universities and accounts for their rankings. Now we will see how this also allows us to predict the evolution of the global education system, from simple to more complex constructs (from a few schools, or channels, to many across the landscape). To do this, we will employ the same type of proof that allows us to predict the size and distributions of channels in a river basin, trees in a forest, or cities in a country or on a continent. All hinge on the prediction that as flow systems become larger, covering a bigger area, they should facilitate the access for the currents that move through them. For universities, this means that a hierarchical vasculature should emerge that facilitates the flow of ideas.

Here is how to use flow geography to predict the linear-logarithmic trend visible in Figure 41 (the same trend would appear fuzzier but still linear if Figure 47 were replotted in log-log coordinates). Imagine an area element *A*1 with a population *N*1. The inhabitants produce things (students, agricultural products, timber, game, minerals, etc.), the flow rates of which are proportional to *A*1. These flow rates sustain a human settlement located on *A*1, where the number of inhabitants is *N*1 and the production is of a different sort (education, knowledge, services, devices). There is a balance between what flows from the area *A*1 to the human concentration *N*1, and what flows from *N*1 to *A*1. The key idea is that both classes of flow rates (area-to-point and point-to-area) are proportional to *A*1, and this means that the size of the human settlement *N*1 is proportional to *A*1.

One type of service that flows from the human concentration *N*1 to the humanity spread over *A*1 is education, educated individuals, books, knowledge, and science. The human settlement in this case is the university, and the area *A*1 is the territory that the university serves. The constellation of universities on the landscape is a reflection of the area constructs of land-city counterflows that cover the entire globe.

As we saw in chapter 7 with our discussion of the Atlanta airport and the evolution of urban transport systems, if the objective is access (a shorter travel time), then the distribution of human movement on the Earth's surface can be viewed as the compounding of area constructs, as shown schematically at the top of Figure 41. Like an area element in a river basin, which feeds the big stream that leaves the area, each area construct sustains the flows that reach a human concentration on the boundary of the construct. It follows that the human concentration on the boundary is proportional to the size of the construct. If the human concentration represents the university, then the university (flow of ideas, impact) is proportional to the size of the area construct that it serves. Over time, the landscape is covered by more and more universities that should have multiple sizes and are assembled hierarchically.

The construction sequence made in Figure 41 is based on area doubling. This construction is how we discover theoretically the pattern hidden in the present-day rankings. Note that the construction of Figure 41 (see page 169) is not a "time sequence" (from small to large) of how the landscape might have been covered by the flowing tapestry of knowledge in history; it is simply a mental viewing of how the patches of the quilt are pieced together. The construction is shown in the bottom left of Figure 41, where the size of the black dot is meant to indicate the rank (that is, the flow rate of knowledge) that the human settlement generates. Given an area, the top-ranked university serves not only the area but also the lesser-ranked universities that are spread on that area.

The bottom left of Figure 41 shows the distribution of multi-rank universities on the landscape after deleting the construction lines used earlier. The hierarchy of ranks is evident: one top university, two universities tied for places 2 and 3, four universities tied for places 4 through 7, and so on. This pattern is discovered here based on pure theory and is represented by the same stepped line as for city sizes in Figure 41. The slope of this line is $-1/2$, in acceptable agreement with what we saw in Figure 48. The important conclusion is not the predicted slope but the fact that the line should be straight and that it has its origin in the area-to-point access for the flow of information between many inhabitants who live on the same landscape. This approach is validated by the fact that we find the same slope in the actual rankings of universities.

Why is the hierarchy rigid?

The short answer is that ideas, science, and education flow all over the globe like water in all the river basins. When numerous researchers value and use an author's work, the idea flows from the author to the user. It flows "well" because of the long history and entrenched geography of the flow network, which are due to the evolutionary process that brought the whole world of information sharing to the present level of effectiveness. The success of this evolutionary process goes unnoticed. And yet, it is the reason the user from one end of the globe actually *looks for*, *finds*, and *trusts* the ideas and young professors produced by a famed university or a professor located at the other end of the globe.

There are many intermediary channels along each route: other universities, disciples of known professors, journals, books, libraries, etc. The intermediaries have evolved into a hierarchical flow structure—the right sizes, put in the right places, nourishing and sustaining each other. Each route is a vascular point-to-area flow (from one source to the entire globe) or a vascular area-to-point flow (from the entire globe to the famed source).

These hierarchical flow designs serve all the scholars well. A hierarchical design that concentrates leading scholars in certain schools is a more effective design for facilitating the flow of ideas than a design that spreads these bright lights evenly across all the schools. Intuitively, we understand that shared resources and the ability to bounce ideas off colleagues should help spread the flow of knowledge. Just as a river basin needs a few large channels and many small ones, so, too, does the river basin of education. The design that has evolved is much older and more polished than a new design that someone may promise to put in place today. The highly ranked and the lowly ranked go together. The flow of science improves in time because each university improves while maintaining the place that it has earned in the global structure.

University administrators who promise to change the rank of their schools by simply stealing one top name from a highly ranked school are defeated by nature every time. Sure, the school's ranking might change a little, moving from thirtieth to twenty-sixth place, but it will still be part of the large group trailing the leaders unless there is a cataclysmic change. The same fate awaits the one who wishes to change rankings by building something artificially big—artificial, because it is not demanded by the natural evolutionary flow and geography that created the tapestry of academic flows that covers our world. An example of artificially big is when a president suddenly decides to spend and build to double the size of his school, because "size matters" in the formula used by *U.S. News & World Report.* Such wishes are analogous to damming, blocking, or digging river channels. The artificial features of the flow network require constant maintenance (spending), more when the artificial does not resemble the natural. In the end, the water knows how and where to flow, the dams break, the dug channels dry up, and the natural design wins.

Age matters in this evolutionary design as it does in all others because it is good for performance. Over time, the river basin improves the positions of its channels, and the channels stay in roughly the same places. The channels have hierarchy: A few large channels flow in harmony with the many small channels. A sudden downpour is served well by the "memory" built into the old riverbeds.

Similarly, the older universities have dug the first channels, which are now some of the largest channels that irrigate the student landscape. Again, "largest" does not mean the greatest number of bodies moving in and out of the classrooms. It means the streams of the most creative, that is, the channels that attract the *individuals* who generate new ideas and who develop disciples who produce and carry new ideas farther on the globe and into the future. The swelling student population is served well by the "memory" built into the education flow structure.

From this view follows the prediction that the hierarchy of universities should not change in significant ways. This hierarchy is as permanent as the hierarchy of channels in a river basin. It is natural because it is demanded by the entire flow system (the globe) in which huge numbers of individuals want the same thing (knowledge).

Is there a way to change rankings? There is, but it takes time, and the river basin provides the perfect metaphor for it. Cataclysmic change (for example, plate tectonics) in the landscape of flow access is the answer. Likewise, the flow of higher education can be diverted through major changes in the loci of generation of new ideas and channels for the flow of information.

Freedom is good for design. We have seen this many times in the evolution of the flow of knowledge, from the movement of Leonardo da Vinci from sponsor to sponsor to the abrupt transformation of nobodies into famous research universities in the United States right after World War II and again after Sputnik. Then, the cataclysmic change was the freedom that attracted the brain drain from postwar Europe, and, after Sputnik, the enormous jump in funding for fundamental research (that is, basic science). These changes had the effect of instituting a marketplace where the flow of ideas was freer.

Not a richer one, not a bigger one, and certainly not one that was to be used as a generator of profit for ancillary and politically correct projects on campus. No. The way to create true academia on a plot of dirt was by putting up a table with free food called ideas. And the truly creative came, to create.

Once you know to look for it, you recognize the evolving vasculature in a river basin because it is a single flow system. It can be much harder to see this design in complex entities like universities because they are channels for many different flows, all superimposed on each other. Education, after all, covers all forms of the transfer of knowledge. It is a global flow vasculature composed of a very large number of flow trees that connect the few who know with the many who need to know in various fields that include mathematics, biology, business programs, and various other areas of inquiry, all of whose ideas flow through tree-shaped, hierarchical channels. It will come as no surprise that the same school can occupy very different positions in the various hierarchies of education—MIT, for example, is a main channel for engineering, not English literature.

To examine this phenomenon, my student Perry Haynsworth and I extended the constructal view of higher education by examining one of the prominent aspects of modern university life: athletics. Specifically, we asked a question often raised by fans of college basketball: Why does it seem that the same schools battle it out each year in the NCAA basketball tournament? Why are there a few university basketball programs that are always successful, while many more continually struggle? Is there a hierarchy in this most competitive arena as rigid as that found in other natural phenomena?

We predicted that the ranking of college basketball teams should be just as rigid because college basketball, too, owes its existence and robustness to a geographical tapestry of area-to-point and point-to-area flows of multiple sizes. The movement of basketball players from high school to the professional level is a flow with its own architecture. There are over 23,000 high schools in the United States. Practically all have basketball teams. The talent ranges from those who would never dream of playing basketball in college to those who aspire to the National Basketball Association (NBA). Several years ago the NBA instituted a minimum-age rule, requiring players to be nineteen years old before entering the NBA draft. As a result, basketball players are essentially forced to choose a university path to the NBA. There are 330 Division I basketball programs that channel players to 30 NBA teams. The high schools and universities are tributaries to the big river that leads to the NBA.

Figure 49 shows how the top university basketball programs arrange themselves when ranked according to their total number of appearances in the semifinal round (the "Final Four") of the NCAA Tournament. When plotted on a log-log field, the data trace a nearly straight line with a descending slope. This feature is important because it unites all natural flow systems that cover the land (see Figures 40, 41, and 48). This distribution is a characteristic of the organization of all flow systems that morph freely and compete for access on the same finite-size territory.

The ranking of the top teams tells a similar story when the measure is the number of players that each team sent to the NBA from 1949 to 2007. These data are plotted log-log in Figure 50,

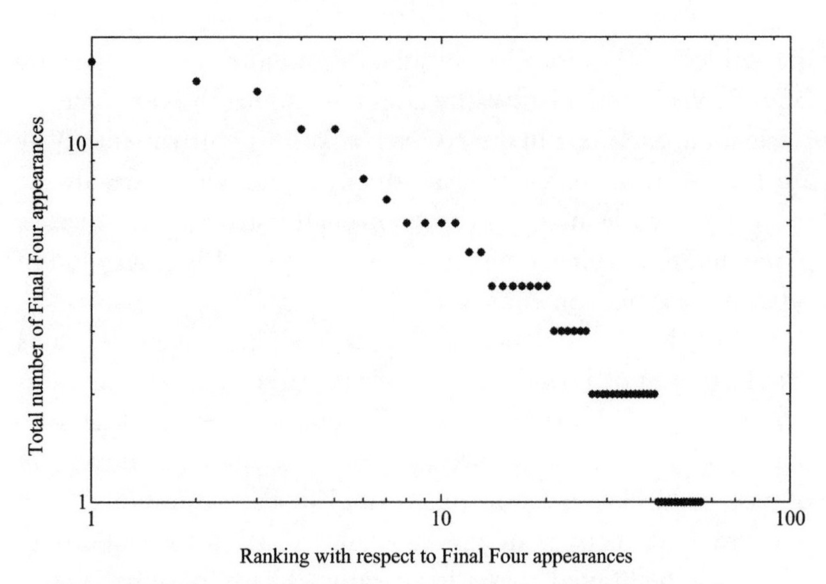

Figure 49. The number of appearances in the Final Four of the NCAA Tournament versus the rank of each team on that list.

and their alignment is the same as in Figure 49. This conclusion is reinforced by Figure 51, in which we cross-plotted the two rankings (Figures 49 and 50) as one abscissa against the other. There is a correlation between success in the Final Four tour-

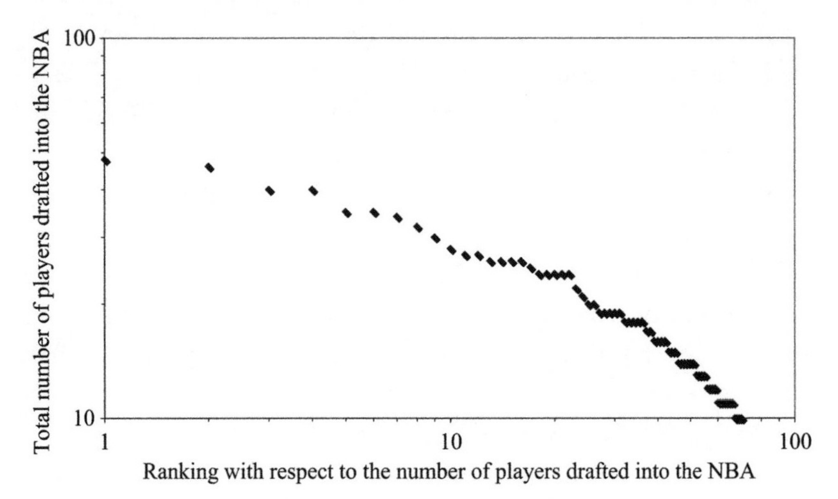

Figure 50. The number of players selected by the NBA from each team versus the rank of the team on that list.

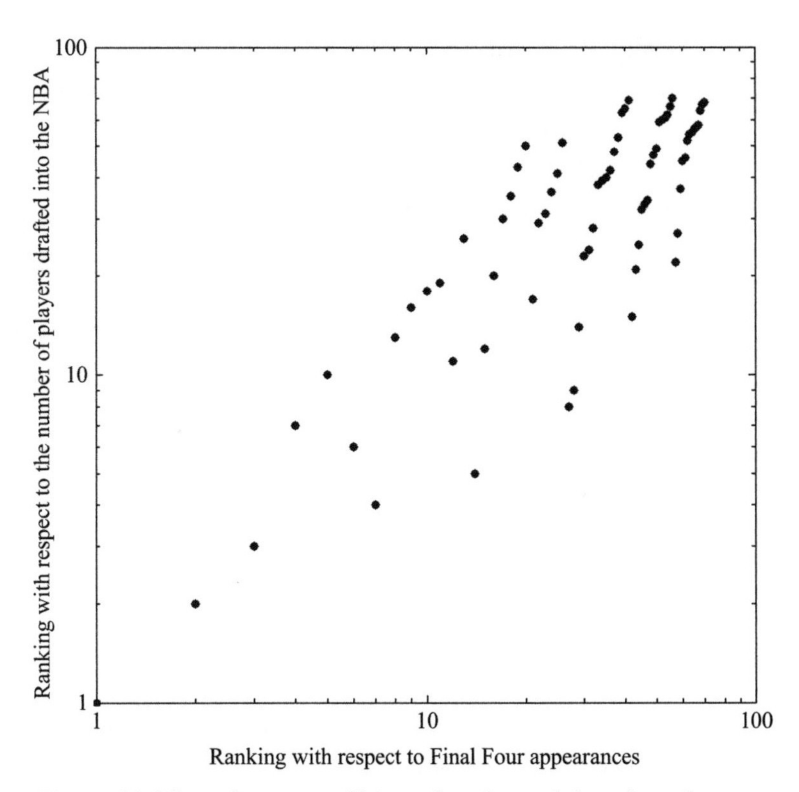

Figure 51. The robustness of hierarchy: the rank based on the number of players recruited by the NBA (Figure 50 abscissa) versus the rank based on Final Four appearances (Figure 49 abscissa).

nament and success in sending players to the NBA. The cloud of data embraces the rising diagonal, and the scatter diminishes greatly in the lower-left corner, that is, at the top of the rankings. The more successful NCAA teams serve as larger and faster streams to the NBA. In conclusion, the hierarchy is rigid.

The robustness exhibited by university and basketball rankings also contradicts the appealing argument that rankings depend on the formula used to calculate the rank. This is explained by the fact that the hierarchy of natural point-to-area flows has two main features: pattern and diversity. These features are evident in the distribution of tree sizes and numbers in the forest and of cities on a continent. They are also present in Figures 47 through

51. The scatter represents the "diversity," which is located primarily in the lower ranks, where there are many competitors for the same rank. It is for this large group that the chosen formula matters, but it matters little; that is, both criteria—the Final Four and the NBA draft—produce very similar results.

To leapfrog a few bicycle racers in the peloton (the thick end of the cloud of data in Figures 48 and 51) is to remain in place, inside the peloton. The runaway racers are well in front, and they have names. Their alignment on the diagonal (Figure 51) represents the "pattern." This is hierarchy, and it transcends all the scheming that goes into ranking formulas and claims that a university (academics or basketball) can be redesigned to score higher in the rankings.

These features (robustness, pattern, and diversity) reinforce the *physics* view that basketball education is a flow system that sweeps the land, while constantly generating flow structures that are more and more efficient. In this evolving design, the top schools are the big branches. They are the few, not the many. Their identity is permanently carved into the geography of the global flow system.

Basketball is just one kind of education that flows with evolving design on the landscape. Every other discipline in which training is pursued by students living on the same area is a flow system with lasting architecture, in which a few large channels flow in harmony with the many smaller channels. The large channels are the highways on which the faster- and farther-moving students travel.

If we superimpose on the global geography all the flow structures of the various disciplines, we begin to imagine how universities constitute their natural global design. Consider now the comparison between the ranking of universities and the ranking of basketball programs (Figure 52). There is no relation between the two rankings. Had they been related, their data would have fallen near the rising diagonal. Most of the universities appear in only one of the rankings. This is why most universities fall on

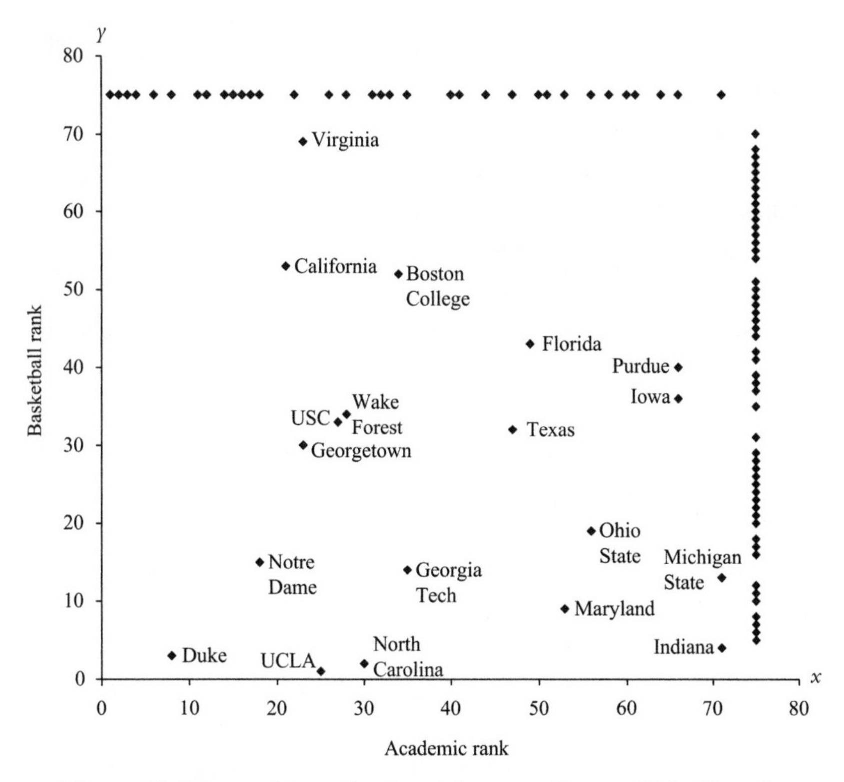

Figure 52. The ranking of universities according to *U.S. News & World Report* (x) versus the ranking according to the number of players drafted into the NBA (y). Most of the universities fall outside the 75 × 75 area: They are plotted on the sidelines (those with $x > 75$ are plotted at $x = 75$, and those with $y > 75$ are plotted at $y = 75$).

the sidelines of Figure 52. They separate themselves into two different worlds, two distinct flow systems on the globe.

When educators and sports announcers refer to college players as "scholar athletes," they misrepresent both worlds. "Basketball students" is a more accurate name, as is "engineering students" for those who study engineering. This stresses again the notion that the global flow of education is a superposition of evolving vasculatures associated with the various disciplines, just as the grid pattern of city streets is the superposition of various flows to various points of interest.

The channels of basketball excellence are not the same as the channels of excellence in academia. The two flow architectures have different histories, memories, and channels. This dissonance is *physics*, and it is worth contemplating because it runs against one of the pillars of modern education: *mens sana in corpore sano* ("a healthy mind in a healthy body"; from Juvenal's *Satires*). Modern education has been right to adopt this doctrine, because it works. This doctrine, however, is not happening by itself, as the two-world reality of Figure 52 demonstrates. The university design needs constant improvements to tolerate and maintain this doctrine; it needs reminders and reinforcement, just like the dams that protect the city from the big river that passes through it.

Our look at university and college basketball rankings underscores two insights derived from the constructal law. First, all natural flow structures that are free to evolve—from the rankings of schools or teams to the size and distribution of channels in a river basin or trees in a forest—are characterized by rigid hierarchies. Second, when we plot these multiscale designs on a log-log graph, we should find a rigid distribution line.

While all flow structures are improving, some are hidden from view as they morph. In social dynamics, the hidden constitute a field of study called "dark networks" and "mafias." Until now we have examined social systems whose flows are relatively easy to recognize and are, on the whole, meritocratic. The selection of athletes discussed in chapter 4 is based on an obvious criterion: their speed on the track or in the pool. The school rankings we've described in this chapter also provided the expected results: Those with greater academic impact or success on the basketball court enjoy greater prestige in their domain. But we all know the world doesn't always work this way. For many flow systems, access to the best channels is based on personal connections, on whom you know and who needs you for the safety and perpetuation of the network. I explored dark networks in my paper "Two Hierarchies in Science: The Flow of Free Ideas and the Academy."

Briefly, I started with the assumption that the membership of the National Academy of Engineering should align with the list of most highly cited researchers. That is not what I found. The resulting comparison—between 171 highly cited authors and 2243 academicians—had a ratio of 1:13. Furthermore, only one-third (60) of the highly cited individuals are also in the Academy, and they represent only 2.7 percent of the Academy membership.

Thus we see that the pattern of generation of good ideas is in disaccord with the pattern of admission to the Academy. The reason is that knowledge and Academy membership are two very different flow systems in the same landscape. The first concerns the flow of ideas; the second, the flow of people already in the Academy.

This phenomenon is prevalent in human relations. It's no accident that the phrase "it's not what you know but who you know" is one of our most enduring clichés. Businesses, for example, are flow systems for goods and services. But they are also vehicles through which owners and managers reward family members and friends with jobs and money. This hiring strategy offers many advantages to the business, especially as it reduces the time spent finding loyal employees. And as long as the company does not become weighed down by mediocrity, it may flourish. This strategy is also a vestige of our feudal past, when the names of the insiders are known to everybody, like the names of the few powerful families in a certain area. Once inside the house, the family invites in the relatives, not the strangers.

Finally, though the constructal law focuses on construction and the coalescence of entities (whether they be raindrops or people) into larger flow systems, the individual remains important.

In my paper "Constructal Self-Organization of Research: Empire Building Versus the Individual Investigator," I noted that empire building is a phenomenon that dominates today's research landscape. Large groups, national priorities (for example, nanotechnology, fuel cells), and research centers dwarf the spontaneous individual investigators. Administrators and the thirst for higher rankings encourage this trend. Yet the individuals do not disappear. The paper explained this by linking the emergence of the large group to the pursuit of greater visibility for the institution as a whole. The visibility (*V*) was modeled as a product of the production (*P*) of ideas in the institution, and the support (*S*) that the institution secures for the production of ideas.

I showed that the coalescence of some investigators into a large group tends to increase *S* and decrease *P*. On the other hand, an increase in the number of individual investigators has the opposite effect. From this trade-off emerges the main and well-known features of the contemporary research organization: the proportionality between the size of the large group and the size of the entire institution, the strong relationship between the visibility of an institution and its size, and the fact that large groups (empires) occurred first in the largest and most research-intensive institutions. I also showed that as the incentives for large-group research become stronger, smaller and smaller institutions find it beneficial to abandon the individual investigator mode and seek a balance between research empires and individual investigators. Thus, the individual researcher will not disappear, for the same reason that older types of movement and ancient animals are not always replaced by newer designs. The invention of carts and automobiles did not spell the end of walking, because that is still a good way to move in many circumstances. Similarly, insects were not replaced by birds, because global flow is enhanced by components of varying sizes. The tendency toward hierarchical organization is not a push toward large, entrenched structures. It is a balancing act in which the few large and many small work together to enhance flow. It takes all sizes.

I know this firsthand. Before my "empires versus individuals" article, I thought I was alone, an anachronism in the eyes of progressive administrators. I was wrong. After this article, I was stopped on campus and contacted by colleagues from around the world who see the world of ideas the way I do. No, the individual is not disappearing; far from it. The individual is everywhere.

CHAPTER 9

The Golden Ratio, Vision, Cognition, and Culture

It has entranced thinkers for centuries, graced with a series of names that suggest its mystic power: the golden ratio, golden proportion, golden number, and golden mean. Those who felt that "golden" didn't do it justice preferred to call it the divine section and divine proportion. Scholars long believed that the Egyptians used it to guide the construction of the Pyramids and that the architecture of ancient Athens was based on it. Fictional Harvard symbologist Robert Langdon tried to unravel its mysteries in the novel *The Da Vinci Code.*

That is quite a feat for a simple problem first described in Euclid's thirteen-volume masterwork, *Elements*. There the Greek geometrist (325–265 BCE) wrote: "A straight line is said to have been cut into extreme and mean ratio when, as the whole line is to the greater segment, so is the greater one to the lesser." We illustrate this by marking (or scribing) the point C on the line AB so that the ratio of the length of AB (the whole line) to AC (the greater segment) is the same as the ratio of the length AC to CB (the lesser segment). Using simple algebra we can compute the ratio between AC and CB as 1.618:1 (or, roughly, 3:2)—a proportion mathematicians represent through the symbol ϕ (phi). Like pi (π), ϕ is what is called an irrational number; it cannot be expressed as a fraction x/y where x and y are integers. Instead it just goes on and on and on; one researcher computed it to 10 million decimal places.

Countless generations have been transfixed by the golden ratio. Until now there has been no scientific basis for explaining its appeal. Through the constructal law we can predict why people should be attracted to shapes with length-to-height ratios close to 3:2. And we learn much more. Indeed, the true mystery of the golden ratio is that it reflects the surprising fact that vision, cognition, and locomotion are features of a single design for movement of animal mass with easier and easier access in time, all over the globe.

Here's how.

The golden ratio was a part of my life before I even knew its name. Since childhood, I was trained to make my drawings look almost square, a little wider than tall. The drawing paper and canvas given to me in art school when I was ten were a rectangular surface with the horizontal dimension (*L*) larger than the vertical dimension (*H*) (Figure 53, left). Ten years

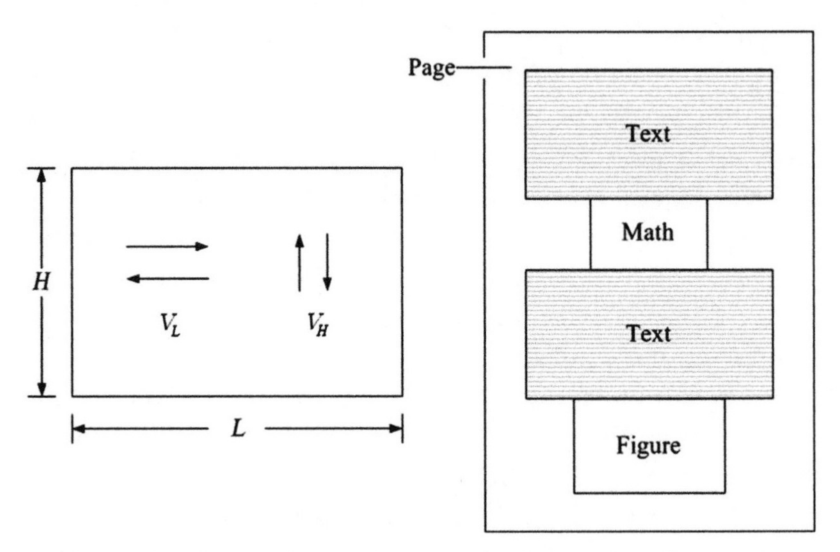

Figure 53. The prevalence of rectangular designs shaped approximately as $L/H \sim 3/2$. The right side shows a layout that breathes and flows: text with graphics and math on a book page.

later, in engineering school, the drawing board was oriented the same way.

Publishing-house artists have advised me that tall figures clash with the horizontal text, and this clash makes both unattractive. Looking around I see the $L \gtrsim H$ everywhere. It is in the shapes of the figures on the page, and in the shapes of the paragraphs—the blocks of text that we tend to read at a glance because we lack time. The text "breathes" when its paragraphs are not too long, that is, not too tall. A frequent piece of editorial advice to aspiring authors is "use shorter paragraphs."

These lessons are not new. It has long been known that certain shapes "breathe" and "flow" better than others. Shapes mean proportions. This "flow quality" of the drawing is undeniably linked to the beauty that we detect in the image.

The connection between proportions and good looks has generated much discussion in science because of a natural tendency: The shapes that we see in the design of books, paintings, and edifices are approximated by a rectangle shaped such that its L/H is equal to the "golden ratio," or 3:2. The fact that generation after generation favors proportions that resemble the golden ratio has fueled an entire literature and mysticism because ϕ has not been deduced from a physics principle.

The race to derive Euclid's ϕ value from principle is justified but misdirected. It is justified because the proportions that resemble ϕ occur around us in very large numbers. This means that the emergence of such designs is a natural phenomenon. A natural phenomenon obeys the laws of nature, that is, the laws of physics. Faced with an unexplained phenomenon, the scientist strives to explain the phenomenon based on known principles.

This effort is misdirected because the physics phenomenon is not ϕ itself. No one has found and measured ϕ on a macroscopic object in nature (ϕ is not like π, which is measurable by dividing the circumference of a circle by the diameter). The physics phenomenon is *the emergence of shapes that resemble* ϕ.

Seen constructally, shapes that resemble the golden ratio arise naturally; they just happen, because human beings are drawn to and create images that incorporate it, including index cards and highway signs, the frames of paintings, cinema screens, and photo prints. The natural phenomenon is the tendency of such shapes to emerge in the devices and artifacts with which we surround ourselves for the same reason that vascular, hierarchical designs and road systems emerge and evolve in the area-to-point and point-to-area flows that crisscross the Earth: because they facilitate our flow access.

Common Designs of Rectangular Images That Resemble the Golden Ratio

Design	$L \times H$	L/H
35 mm film	36×24 mm^2	1.50
Computer display	1024×768 mm^2	1.33
Canon 5D	4368×2912 mm^2	1.50
Canon S3 IS	2816×2112 mm^2	1.33
HDTV	16×9 in^2	1.80
Photographs	6×4 in^2	1.50
	7×5 in^2	1.40
	10×8 in^2	1.25

As with all evolving flow systems, we ask two key questions: What is flowing? How does the design facilitate that movement? Regarding the golden ratio, we ask: What flows when we look at a page with text, math, and art? And why do the shapes with $L/H \sim 3/2$ appear to "breathe" and "flow" better than the others?

The answer is that information flows, as images, from the page to the brain (Bejan, 2009b). Through the constructal law, we understand that information is the current that moves through channels that evolve to provide greater access to these currents. The flow of information involves a wide array of mating streams. We have already seen how science evolves to transform mountains of observations into principles so that its knowledge can spread more easily over larger areas. We have witnessed the same phenomena in the design of universities and the Internet. Now we are looking at an even more fundamental component of this global flow—the movement of information from our line of vision to our brains. Using the constructal law, we predict that every aspect of this flow system, from the design of information that enters our eyes to its movement through our brains, should morph to increase flow access naturally. The evidence in support of this principle is massive.

The architecture of the brain consists of bundles and bundles of constantly forming and adjusting tree-shaped channels of neural fibers that provide easier and easier point-to-volume and point-to-area access to the regions of the brain that control various activities. This also holds for the connection between each elemental volume of the brain and the rest of the brain volume, and vice versa. The visual sensors and nerves in the retina are configured in order to provide greater access between one surface (the retina) and one point (the optic nerve).

The external architecture of this flow system has also been morphing in the constructal-law direction to generate a flow configuration in time, toward easier flowing. Here again the evidence is massive. The evolution of writing, toward simplicity and universality (one alphabet), is one phenomenon of design generation. The evolution of spoken languages, especially the emergence of lingua francas, from ancient Greek and Latin to French and now English, is another example. That is, just as the hardwired channels in the brain have morphed over millions of years to provide greater access to larger streams of information, so, too, have the channels created by humans to spread those currents over wider areas. The evolution of book design, library design, currency design, photography, eyeglasses, dashboard, and computer screen design is the same phenomenon of facilitating the flow of information between the page and the brain.

This brings us to our proof of the constructal design of the golden ratio—which stems from the general design principle detailed in chapter 7. We saw there that when two types of movement are at work, an efficient system balances the time to move slow and short with the time to move fast and long.

Start with the area $H \times L$ shown in Figure 53 (left). The shape of the image is part of the architecture of the information flow system, and it is free to change, just as designers were free to change the shape of the rectangle that is the Atlanta airport. The constructal law predicts that the shape that emerges should allow the eye to scan the rectangular area $H \times L$ with the greatest ease, that is, in the shortest time.

In the simplest description, "to scan" is to sweep the image completely, once horizontally and once vertically. The horizontal sweep covers the length, L, with the average speed, V_L. The horizontal sweep time is $t_L = L/V_L$. The vertical sweep covers the distance, H, with the averaged speed, V_H, and time $t_H = H/V_H$. Combining these two equations, the total time required to scan the image is of order $t = L/V_L + H/V_H$.

The area of the image (A) is fixed ($A = HL$), but the shape of the image (L/H) can vary; given freedom, it can evolve to resemble any shape we choose. The total time is $t = L/V_L + A/(LV_H)$, and it is minimal when $L = (AV_L/V_H)^{1/2}$, which represents this rectangular shape:

$$\frac{L}{H} = \frac{V_L}{V_H}$$

The first implication of this result is that the shape of the image influences how it is perceived, understood, and recorded. At this stage in the analysis, V_L and V_H are not known, and neither is L/H.

The second implication is that when the image is shaped according to the above equation, the horizontal sweep takes just as long as the vertical sweep, $t_L = t_H$.

As we have seen, the balancing of two different flow regimes is a common design feature for flow access (for example, it is found in the design of city traffic, river basins, and lungs). In the present case, $t_L = t_H$ means the time to scan long and fast must be the same as the time to scan short and slow.

The third implication tells us the broad outline of the shape of the rectangle we are designing. In it, L must be greater than H because, as we will show next, V_L is greater than V_H. This is because we scan things more quickly on the horizontal than on the vertical. The reader can test this: Since human eyes are side by side, it's easy for you to scan horizontally, while to scan vertically triggers the urge to tip your head.

To predict a more precise shape for our rectangle, we must also consider the organ that is scanning the image. It is quite telling that the eye mechanics literature contains information on horizontal eye movement (V_L) but not on vertical movement (V_H). This record is important because it underscores the fact that we generally perceive the world as a roughly horizontal tableau. Our world is flat. Our supply of images reflects the orientation of the landscape. Danger generally came to prehistoric humans from the sides and from behind, not from above or below.

Like the L/H ratio predicted above, the positioning of our two eyes on a horizontal axis is a constructal-law design feature. The horizontal orientation of the eye-eye axis has emerged because it facilitates the flow of visual information from our horizontal environment to the brain.

The horizontal shape of our field of vision is approximated by the construction shown in Figure 54. The length scale of the disk (R) exists because of the distance between the eyes. The size of R is not the issue—the existence of R is part of the constructal design of how we see the world. The superposition of the two disks is the binocular area that we can cover with them. If one eye sweeps one disk horizontally (length $2R$, time t_L) and vertically (length $2R$, time $t_H = t_L$), then, because of the superposition of the two disks, the horizontal length scanned by the two eyes is $3R$. The horizontal and vertical speeds are $V_L = 3R/t_L$ and $V_H = 2R/t_H$, and because $t_L = t_H$, the ratio of speeds is $V_L/V_H = 3/2$.

The binocular area can be approximated by superimposing on

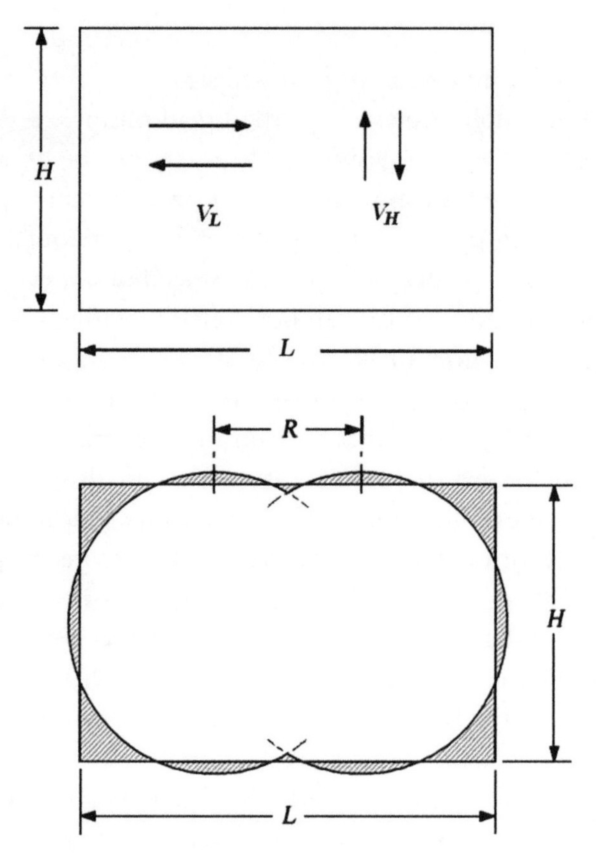

Figure 54. The closest rectangular approximation of the binocular area has the shape $L/H \sim 1.47$.

it a rectangle with the right shape and size. Two approximations are closely represented by the same rectangle in the lower part of Figure 54. The first is the rectangle that mimics most closely the curvilinear contour of the binocular area: The sum of the areas formed between the rectangular and the curvilinear contours is minimal. The dimensions of this rectangle are $L = 2.768\ R$ and $H = 1.876\ R$, and its shape is $L/H = 1.475$. If, in addition to minimizing the total area of mismatch between the two shapes, the rectangular area must be equal to the binocular area, the best rectangular shape is $L = 2.724\ R$, $H = 1.856\ R$, and

$L/H = 1.468$. This second shape is practically the same as the first.

Humans scan the world on a two-dimensional screen approximated by a rectangle with the shape $L/H \sim 3/2$. We scan the long dimension faster than the vertical dimension, in such a way that to scan long and fast (L, V_L) takes the same time as to scan short and slow (H, V_H). This is the best flowing configuration for images from plane to brain, and it manifests itself frequently in human-made shapes that give the impression that they were "designed" according to the golden ratio (see the table on page 224).

This principle-based explanation has several broad implications. First, it unites two seemingly disparate phenomena: biological and cultural/technological evolution. Both are governed by the constructal law. Vision has evolved over hundreds of millions of years to enable biological creatures to scan the world faster and more efficiently. Through the much shorter annals of human history, our designs have evolved to create easier access for the flow of information to our brains, and to humanity at large. To take one recent example, old computer and TV screens had aspect ratios close to 1.33; this was a first-cut design. However, as new technologies loosen design restrictions, the screens have morphed toward wider shapes, with L/H values closer to 3/2. The future will bring more designs that resemble this.

In addition, this discovery offers new insights into how and why human beings prize harmony and balance. The sublime beauty people find in objects that resemble the golden ratio is not due to some abstract quality that only the finest, most aesthetically attuned minds can appreciate. We consider them lovely and intriguing because they are in tune with how we see the world and are therefore useful. If there is a "mystical," timeless secret to the golden ratio, it is the fact that it connects humanity to nature: Everything that flows (including us) generates designs that enable it to move more mass, more easily, on Earth. For humanity, this physics phenomenon brings not only actions and movement, but pleasure; thus, when we see entities that help us achieve this better movement (longer life, etc.), we find them pleasing to the eye and we make more of them. To paraphrase the poet John Keats, "Beauty is movement, movement beauty."

This phenomenon leads us to the most important idea that springs from our constructal explanation of the golden ratio phenomenon: the integrative design of the movement of biological mass on Earth and the rise of cognition. To appreciate this we must keep in mind that the constructal law is a principle of physics that governs the evolution of all flow systems. Inanimate and animate systems evolve *in order to flow more easily.* In chapter 3, we used this mental viewing to predict the scaling laws of all animal locomotion. We saw why larger animals should be faster and observed the time direction in the evolution of all biological systems—from the first organisms in the sea, to the rise of animals on land, then to those in the air. Designs evolved so that at each subsequent stage of evolution the newer animal forms were able to cover more area for less consumption of useful energy.

This follows from the subsequent argument: The constructal design of animal locomotion calls for a balance between the work of lifting mass (W_1) and the work of moving the body horizontally against the resisting medium (W_2). Because of the balance between W_1 and W_2, the total work $W_1 + W_2$ is of the same order as W_1 or W_2, where W_2 is the drag force times the distance traveled.

For fliers, the drag force is $\rho_a L_b^2 V^2$, where the body length scale L_b is $(M/\rho)^{1/3}$. The spent power is the drag force times V, that is, $\rho_a L_b^2 V^3$. The work spent (W) during flying to the distance, L, is equal to the spent power times the travel time, L/V, where $V \sim (\rho/\rho_a)^{1/3} g^{1/2} \rho^{-1/6} M^{1/6}$, cf. Figure 23A. This derivation yields $W \sim (\rho_a/\rho)^{1/3} MgL$, where $(\rho_a/\rho)^{1/3} \sim 1/10$.

For swimmers, the same derivation yields $W \sim MgL$, which is one order of magnitude greater than for fliers. For runners, the derived work requirement is between fliers and swimmers, $W \sim r^{-1}MgL$, where $1 < r < 10$.

In summary, work requirements decrease and speeds increase in the direction sea → land → air, which is the design evolution direction dictated by the constructal law (Figure 55).

Locomotion design is a manifestation of the constructal law, and it has been improving throughout the big history of biological forms and flow systems on Earth. This is why animal locomotion first emerged in the oceans, spread onto land, and later rose into the air and not the other way around. The time direction of this evolution has been toward higher speeds, and it is shown qualitatively in Figure 56, which is a detail of the side plane of Figure 55. More movement and more mixing of

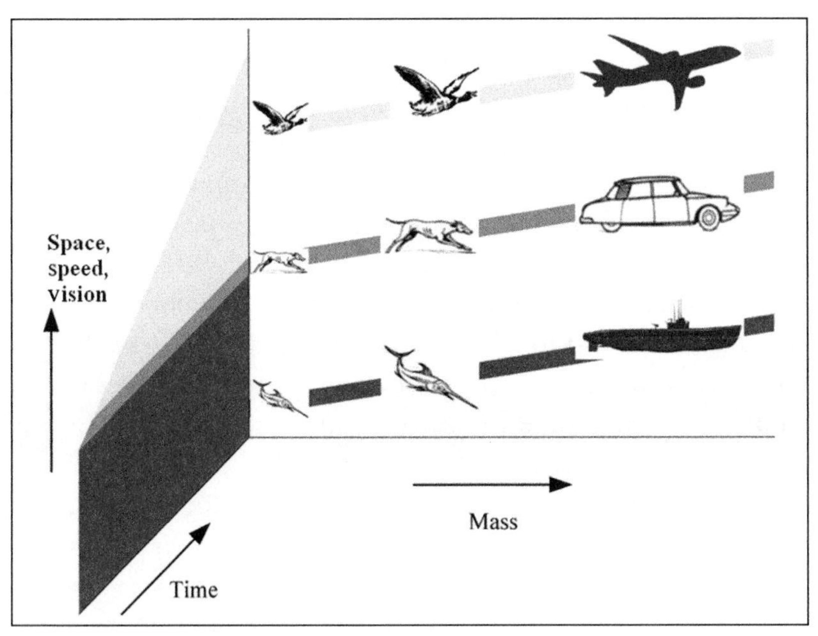

Figure 55. Space, speed, vision, mass, time. At any point in time, the biosphere churns itself with a huge diversity of animate moving bodies organized according to a pattern. The larger bodies tend to have higher speeds, lower frequencies of body movements, and larger forces.

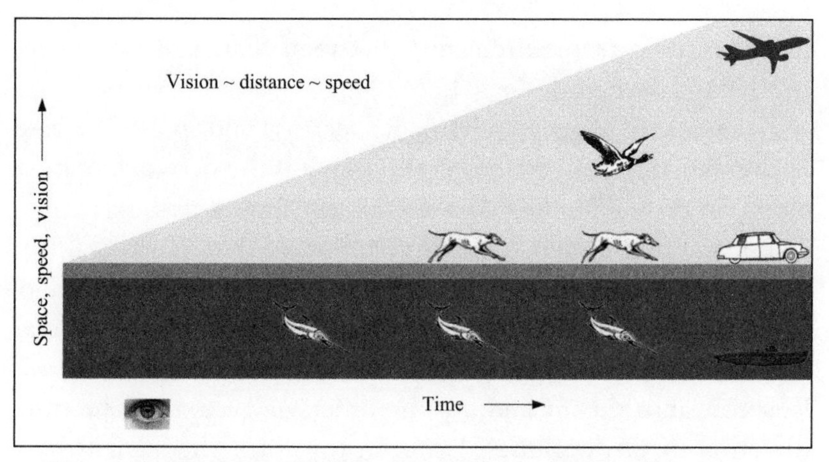

Figure 56. Space, speed, vision, time: The evolution of the biosphere from prehistory to today. Animal flow has been spreading in space, toward higher speeds and the ability to see farther. This montage fits on the left plane of Figure 55.

the Earth (upward in space) have always been aligned with time, more speed, and more space traveled per unit of animal mass and useful energy consumed.

The big jump in the perfecting of the animal locomotion design was the emergence of the organ for vision, the eye. This has made the flow of animal mass much more efficient, faster, and enduring. With vision and cognition—the ability to process and to respond to what we see, hear, and feel—the flow of animal mass designs for itself ceaselessly better channels to flow: straighter, safer, with fewer obstacles and predators. With organs for vision, the animal minimizes danger from ahead and from the sides. This is the link between vision and locomotion, and the fulcrum of the single design of animal movement on Earth.

Animal movement with vision is *guided locomotion.* The evolution toward more movement, space, and speed (Figure 56) is also the evolution toward better, more powerful vision. This is also the evolution of all vehicle technologies. Longer range for a vehicle goes hand in hand with higher speed and the ability to see farther. The bird sees farther than the dog. The jet pilot sees farther than the tank driver.

This step-change in the animal locomotion design is known as the Cambrian explosion (circa 530 million years ago), and its time arrow is in complete accord with the constructal law, toward more space, speed, and mixing of the Earth's crust. The Cambrian explosion encompasses all these advances. The animal design with vision and cognition came after the animals without vision and cognition, not the other way around.

The payoff of this theoretical connection is much bigger than the explanation of the golden ratio. It is the oneness of vision, cognition, and locomotion as the design for the movement of animal mass on Earth. Shapes that resemble the golden ratio facilitate the scanning of images and their transmission through vision organs to the brain. The speeding up of this flow goes hand in hand with the architectures of the nervous system in the eye and the brain. Dendrites facilitate the point-to-volume flow of information inside finite volumes, and the new point-to-volume connections can occur naturally in the brain. The name for this constructal evolution of brain architecture, every minute and every moment, is cognition—the phenomenon of thinking, knowing, and thinking again, better.

"Getting smarter" and the wisdom of the saying "work smarter, not harder" are the constructal law in action, another way to move more mass more easily on Earth. At the end of the day, intelligence and knowledge emerged as internal features of the flow design.

This is also true of culture. "Culture" is a short name for the acquired knowledge passed on through the generations—it is why each fish does not have to discover that it is better to swim in a school and why wolves know to hunt in packs. For humanity, culture is the endless list of flow architectures we have created that cover and sweep the globe. These include all the known and still unknown forms of human movement—walking, working, and staying alive by using and developing enhancements that make life easier: knowledge, shelter, hygiene, language, writing, social organization, music, visual arts, and the running stream of novelties, inventions, and secrets unlocked. We call these good things "ideas."

Good ideas travel and persist. They keep on traveling. This is why culture is a constructal design—a tapestry of morphing linkages in our minds and on the globe—all superimposed on the same area (the globe) and in the same volume (the brain). As such, culture is the same kind of design as the tapestry of vascular architectures, animate and inanimate, all superimposed on the Earth's surface.

The brain counterparts of the observed (external) flow architectures are precise, like the reflections of images in a highly polished mirror. We "reflect" when we think. We generate mental viewings of the world of movement and actions that surround us. We do this every instant. We reflect not only on the past and the present but also on the future. We are wired with the ability to fast-forward the tape of our mental viewings. We know what will happen when we find the door and walk through it, and what will happen if we walk into the wall. We make choices, constantly and unconsciously. We construct our movements on Earth, our existence, into the future.

The ability to reflect and to run the movie tape forward has been increasing in all of us, humans and animals. The better and more easily we guide and power our movement, the more our movement (our culture) persists. Every living thing possesses this ability, to use the environment as fuel (food) and to guide its movement with sensory organs. Broadly speaking, the evolution of all life-forms has been toward more, easier, faster, farther, and longer life-movement. This is synonymous with evolving toward knowing more and doing more things, toward being smarter.

It makes sense that in addition to the ability to *reflect*, humans have the ability to *speculate*. This means to form mental images of how nature *should be*, without peeking at nature first. To speculate is to look into nothing except the mirror of the mind (*speculum* means "mirror" in Latin). This purely mental activity means "theory," and the ability to theorize, too, is evolving so that, in the final analysis, our movement is made easier.

Culture spreads because humans are on the move. Culture flows from those who have it to those who feel the benefit of acquiring it. Culture flows from high to low, like all the other streams that obey the second law of thermodynamics. What is news or education but information one person possesses and another desires to know? When both parties know the same thing, the information is no longer news and education stops. Similarly, all migration—whether we are speaking of fish, birds, or people—is the movement of empty vessels to places where they might be filled. Although it is politically correct to speak of cultural exchange, this is nonsense. The Romans spread out to acquire what they lacked (slaves and internal security), and the barbarians attacked them at the same time to acquire what they lacked (food, shelter, culture).

History and geography are the established disciplines that teach where we came from and how we got to be smart, self-sufficient, and safe. In a nutshell, we are becoming more and more civilized because culture flows. Interruptions like those that caused the Dark Ages can happen, but the natural tendency expressed through the larger pattern is toward more culture flowing through.

In antiquity the flow was on the backs (and in the heads) of individuals roaming on the landscape. When the roaming individuals were numerous, the effect on the culture of the invaded was cataclysmic. When the invading group possessed more culture than the invaded population, the effect was emancipation and advancement. When the invading group possessed less, the effect was the Dark Ages and Soviet communism. Both happened in the Europe I know.

Culture, as a flow, is much more complex than human locomotion. Culture is inventing and knowing the channels and the ways in which to move. Culture is the knowledge to produce, harness, distribute, and use power. The rule of law—waiting patiently in line, as opposed to triggering a stampede at the gate—is culture. Those who possess it travel far, which is the opposite of being taken to the morgue.

Culture is many hands that work the land, for seeds to be planted. Culture is good for movement. Lack of culture is not. Both realms exist, one in the world of evolving civilization (as written in our history books), the other in the caves. The less cultured understand this, because they are attracted by the obvious effect of culture—more plentiful food, shelter, and, above everything, freedom. They speed walk toward us. Nobody is forcing them to wear suits and speak English.

Freedom is good for design, and design means movement. This dictum follows from the constructal law, because freedom is a prerequisite for the ability to change, to move more easily. It is, by the way, captured in Darwin's hunch that the survivor is the one who adapts.

I did not learn this from Darwin but from my father, Dr. Anghel Bejan, a veterinarian who, during the most murderous period of communism (called the dictatorship of the proletariat, even though the proletariat had no say in it), would pronounce loudly to anybody who would listen: "Look in the eyes of the dog. He is saying to you: Leave me alone. I want to be free."

We should listen to the dog, because without freedom we would have no movement, no culture, no lasting presence on Earth. Freedom is physics.

CHAPTER 10

The Design of History

What is life? How has it evolved? Where is the world heading? Poets, philosophers, scientists, artists, and most people with a spark of curiosity have pondered and debated these questions since time immemorial. The musty stacks in our world's great libraries reflect our unyielding efforts to discover better answers to these eternal questions.

The constructal law enables us to take a significant step in this quest. It teaches us that anything that flows—which is just about everything—is "alive" because it evolves as it flows. Life is the persistent movement, struggle, contortion, and mechanism by which animate *and* inanimate flow systems morph to generate better access for what flows. When the flow stops, the configuration becomes a flow fossil (for example, dry riverbeds, snowflakes, animal skeletons, abandoned technology, and the Pyramids of Egypt).

This view challenges the entrenched line of thought that assigns humanity special standing in the natural world. One iteration, usually based in religion, casts mankind as the apex of divine creation. In science, Darwin and his followers have helped knock us off that pedestal by connecting human beings to "lower" life-forms. But their effort is only a half measure because it assumes that biological systems are fundamentally different from everything else.

The constructal law corrects this. While recognizing important differences—no one should ever confuse people with rivers—it identifies the single principle of physics that makes an entity "alive," that governs the evolution of rivers and rhinos, lightning bolts and lizards.

In chapter 3, we cited Stephen Jay Gould's thought experiment in which he imagined replaying life's tape. His metaphor was apt: Life is a moving picture. His mistake was in suggesting when the film should *start*—perhaps 600 million years ago, when multicellular entities blossomed, or even 3.5 billion years ago, when biological activity began swirling in the primordial soup. That is tantamount to walking into the middle of a movie. In fact, the film truly began running when the universe formed and flowing currents began acquiring evolving designs. The emergence of biological organisms was a wondrous event, but it was not the magic moment when "life" suddenly appeared. It was not the start of evolution but a plot twist in the larger story of mass and energy flows being shaped by the constructal law. Life—flow, with freely morphing configuration—was there from the start.

By teaching us that life is flow, the constructal law collapses the false distinctions between the animate and the inanimate, providing a single, universal law that accounts for all design and evolution in nature. The constructal law shows us that humanity does not stand apart from nature but is a manifestation of, and governed by, nature. In fact, everything on Earth is a manifestation of nature; nothing is "unnatural" or "artificial." Even the "nonnaturally" arising chemicals and inventions humans create reflect the natural tendency to make designs that allow us to move our mass faster, farther, and over longer lifetimes. The constructal law reveals that history is not a series of discrete narratives—the story of rocks or rivers or plants or people—but a single story woven together from the various flows that mate and morph on Earth.

In this final chapter, we will pull together all that we have discovered to write a new history, a constructal history, of life on Earth. We will do this in two parts. The first underscores the oneness of nature by focusing on the source of almost all movement, all life, on Earth—the sun. For all the diversity we find in nature, the history of our planet is, in fact, the unfolding story of the interaction between solar energy and the mass it sets in motion. This fundamental view is powerful because it allows us to develop a new, concise, and all-encompassing picture of the global design of nature. Life on Earth is a tapestry of engines (which drive every flowing current) and brakes (all the resistances and losses that the currents encounter). All these designs, the engines and their brakes, evolve hand in glove and are governed by the laws of physics that now include the constructal law.

In the second part we use this engines-and-brakes design to see that the evolution of life on Earth began with the emergence of inanimate designs, continued with the rise of animate designs, and progressed to the appearance and evolution of the human-and-machine species. In the process, we deliver on the promise made in the introduction: to see the world anew, as it really is, constructally.

We start with the biggest flow system that surrounds us: the Earth, which is nature itself. Nature looks complicated, all the more so as we separate its components into walled-off areas for study—the atmosphere here, the hydrosphere there; the lithosphere in one room, the biosphere in another. Nature is in fact a tapestry woven on a very simple loom. The designs in all these spheres consist of many flow types and sizes, all governed by a law of physics. Even better, the tapestry itself—the single design created by all the morphing and mating flows on Earth—is constituted according to the same law. The designs of nature are not random or haphazard. All designs fit—the animate and the inanimate, the small and the large, the human and the not human. They do not fit perfectly and never will. However, they ceaselessly tend to fit better and better over time. The fact that everything that moves is free to morph means that *every thread* and motif of the tapestry evolves so that the *whole* flows better.

The more we rise above the details, the simpler the tapestry design becomes. Taking a bird's-eye view is very good medicine for those sickened by the dogma that nature is complicated, diverse, random, nondeterministic, complex, emerging, fractal, turbulent, nonlinear, chaotic . . . words that sound scientific but all mean one thing: "I cannot predict, therefore I give up."

I was taught the bird's-eye view method at MIT by my famous professor of dynamics, J. P. Den Hartog. He was an artist of the simple, in a discipline that was already cluttered with immensely complex mechanisms (and that was decades before the blur of computer-generated simulations of "anything" today). He urged his students to step back, look at the whole, make it simple but "do not throw the baby away with the bathwater." Professor Den Hartog was teaching the art of seeing the essential.

The race to explain design in nature has been hindered by the search for answers at infinitesimal scales. This has blinded us to the fact that design in nature is decidedly a macroscopic phenomenon, not a microscopic one. Design is what we see, what we imagine, what keeps us awake. It is not the discrete particles and probabilities that animate the crowd that marches today. It is the coming together of larger and larger quantities of mass. It is the macroscopic black lines that emerge and evolve on the macroscopic white pages of nature, the visible shapes and structures all around us.

Here is how the fundamental principle, the design of nature, jumps at us if we take a bird's-eye view of the whole big bag of components and fine details on Earth. The sun shoots streams of energy in all directions. Some of these streams are intercepted and absorbed by the Earth. Altogether, they represent one current of energy that flows out of the sun and into the Earth. This current flows from sun to Earth because the sun's temperature is higher than the Earth's. Similarly, a current of energy flows from the Earth to the sky, because the Earth is warmer than the sky.

Because solar energy heats the Earth unevenly, the heat on Earth flows in accordance with the second law of thermodynamics (from hot to cold) and the constructal law (with evolving design). One word for this constantly morphing design is *climate*, and, not surprisingly, the main features of the Earth's climate (climate zones, temperatures, wind speed, and so on) have been predicted from the constructal law. This fundamental finding belies the claim that we need impossibly complex models to predict the Earth's climate.

The constructal law takes this observation further. It accounts for the fact that all the live systems on Earth (not just the climate) intercept and use this energy from the sun. As a result, the entire Earth is flowing, especially in its spherical shells that house the designs that we observe and interest us—the hydrosphere, atmosphere, lithosphere, and biosphere. They all flow by acquiring configurations that evolve in time. How do these guts fit in the big animal? The animal is the globe, as an intermediate stop for a train of energy from the hot sun to the cold sky. Just as the design of animals has evolved so that they can move their mass farther per unit of fuel consumed, all the flows on Earth have evolved together to facilitate global flow.

Anything that moves on Earth does so because it is driven. The driving is done by very subtle engines, one engine for each flow. These engines have many names—the atmospheric circulation that brings the snow and the rains on the mountains and plains so that water will flow in the rivers; the solar heat that falls on the warm zones and drives the ocean currents; the muscles and tissue of animals that propel them horizontally across the map, on land, in water, and in air. No matter how numerous and diverse, all these engines are driven by fuel (for example, the food for animals) that comes from the sun in the form of the heat current (Q) intercepted by the Earth. These engines convert fuel into heat to perform work (W). All of them.

In the upper part of Figure 57, we now imagine that all these engines are represented by one engine, which uses the heat input,

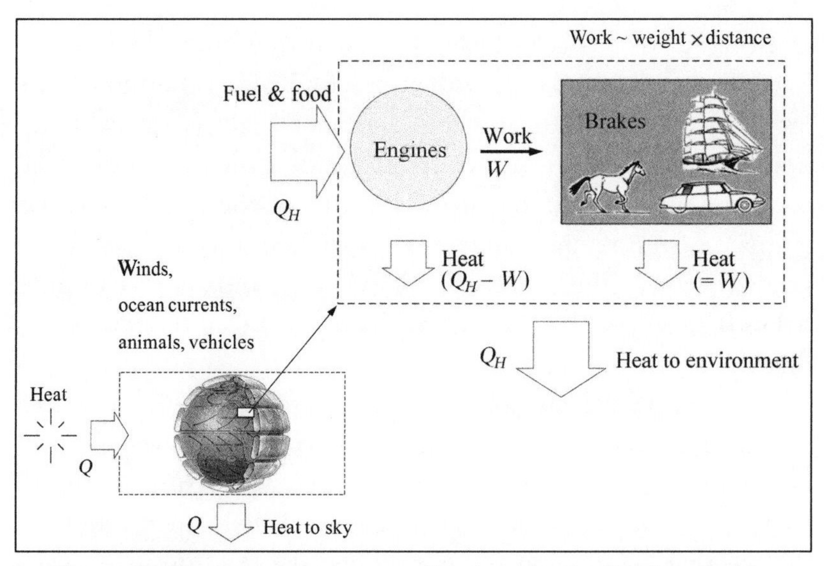

Figure 57. The solar heat current (Q) that hits the Earth and ultimately sinks into the cold universe. The Earth temperature settles at a steady level between the sun temperature and the sky temperature. The upper detail shows the two phenomena that the solar heat current drives as it passes through the Earth. First, Q_H drives flows (natural mechanisms [contrivances] with moving parts) that function as "engines" and produce work, *W*. Second, the work, *W*, is dissipated in the "brakes" that form between these flows and their immediate environments (neighbors). Seen as a whole, the flowing Earth (engines and brakes) receives Q from the sun and rejects Q completely to the sky. The whole Earth is an engine-and-brake system, containing innumerably smaller engine-and-brake systems (winds, ocean currents, animals, and human-and-machine species).

Q_H, in order to produce the work, *W*, that is needed for driving (forcing) all the things that move on Earth. The difference between heat flowing in and work flowing out is $Q_H - W$, which is rejected as heat to the environment.

This completes the first part of the story, the subtle part, because we do not see "engines" in what moves around us. The winds and the rivers seem to move *by themselves*. My colleagues in physics and engineering refer to these flows as "free convection" and "natural convection" and "buoyancy driven" flows. They are thought of as free and natural because we do not "pay" for moving them.

Yet they are driven.

The second part of the design of nature is the movement that occurs against resistances that constantly try to stop it. Without such resistances, the objects driven by the work, W, would accelerate forever and spin out of control. This is not how nature is. All the driven things dissipate all the driving W in the brakes that form between the moving objects and their immediate surroundings; the environment of any engine is a brake. (See the shaded box in Figure 57.) These brakes are diverse and include the friction and other forms of resistance that swimmers, runners, and fliers encounter as they move across the landscape (see chapter 3) and the riverbanks that rub against the flowing water.

The "aha!" is that all the work, W, is dissipated into heat (called Q_{diss}, and equal to W), and that Q_{diss} is also rejected into the environment and eventually the cold sky. Altogether, the Earth rejects heat into the sky from the engines ($Q_H - W$) and from the brakes (called Q_{diss}). The sum of these two heat currents is Q_H, because $Q_{\text{diss}} = W$. The conclusion is the same as what we saw in the lower part of Figure 57: The total heat current that is rejected into the sky is the same as the heat current received from the sun.

This statement may seem counterintuitive because it suggests the idea of a free lunch—that is, that all the engines on Earth use energy from the sun to move and yet, in the end, all the energy that reaches the Earth is sent back into the sky. This riddle has two parts. First, remember that in the steady state the Earth cannot store energy. What arrives from the sun penetrates the spheres, sets things in motion, and bounces off into the cold sky.

Second, the solar heat current "sets things in motion" because riding on Q is a stream of useful energy (exergy). This useful energy is what "engines" convert into the work that moves things and which is destroyed completely. Useful energy (exergy) must not be confused with energy.

The moving things that make up the hydrosphere, atmosphere, lithosphere, and biosphere are a tapestry of engines attached to brakes. As the engines we call rivers or animals move, they must rub against the brakes of their surroundings. They must get the environment out of the way. "Environmental impact" is the other name for movement, for the flowing water that carves channels on the plain, for the flowing earth that has created valleys and mountains, for flowing people who have built cities on the landscape. This causes friction and other frictionlike losses, which dissipate their work into heat that is rejected into the cold sky. This dissipation (or destruction) of work is what we really do when we "use energy." In reality, we use food, fuel, and solar energy and in the process destroy their useful-energy content.

This completes the bird's-eye view and confirms the continuity of the heat current (Q) through the Earth, from the sun to the sky.

The only reality we know is the one we see before our eyes. So, in the largest sense, we do not know whether the engines that drive the Earth's flows are very efficient (like the ones in Carnot's imagination) or very poor. We don't know, and, fortunately, this is not the issue. What we do know is that flow resistances plague all of them, that they are all imperfect. We also know that they all have the tendency to acquire evolving designs that distribute these imperfections so that they flow more easily. It is this tendency that is the constructal law, and it can be restated in different ways depending on which side of the engines-and-brakes model we focus on.

First, if the flows and moving parts of an engine morph in time so that they move more easily, the engine design evolves in the direction of producing more and more work (W) from the fixed heat input (Q_H). This is the direction of improvements in efficiencies—animal designs that are better fit for moving more animal mass on Earth and geophysical currents that move more water and air mass through hierarchical, vascular designs (Figure 58). The constructal evolution of these engines

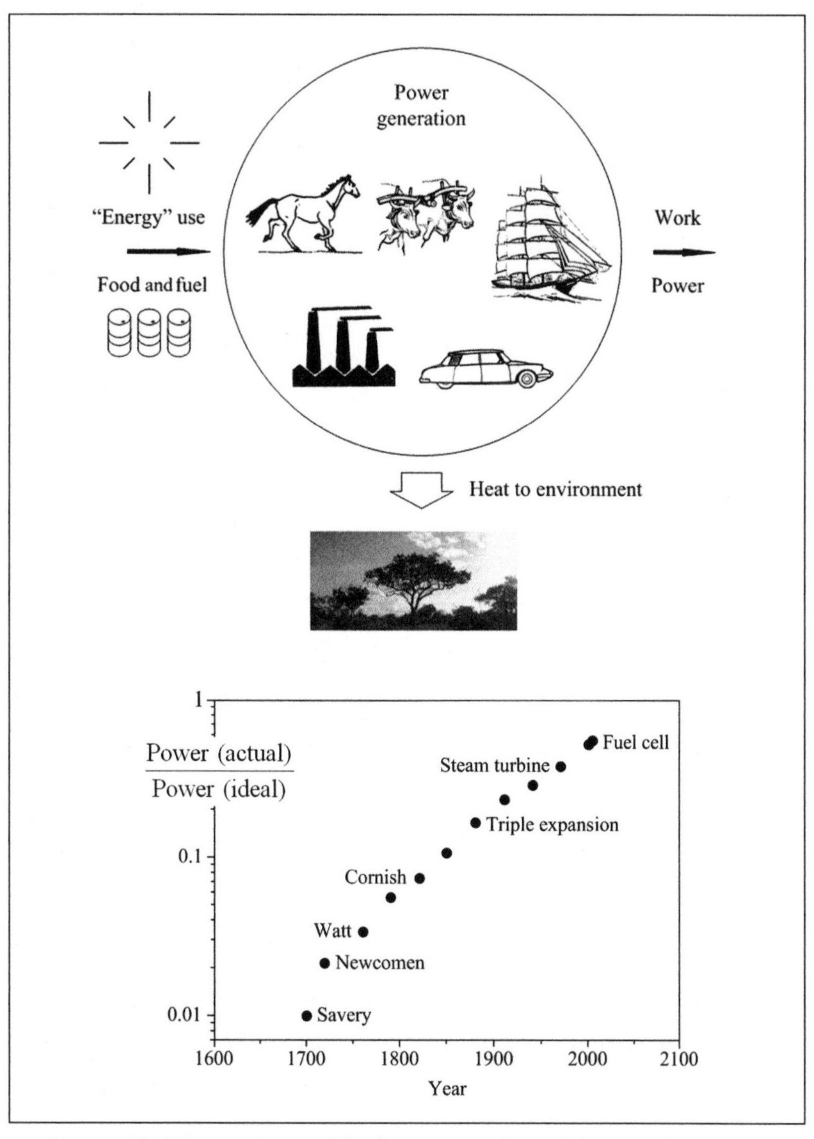

Figure 58. The engines of the human-and-machine species are the power-generation technology, which has always been evolving. New power-producing configurations are added to existing configurations, and the efficiencies of the new surpass the efficiencies of the old. We are all human-and-machine species, each a little different from our neighbor, each evolving right now as a flow system on the globe, as part of the global tapestry of all the human-and-machine species like you and me.

is amply documented in the history of technology. For example, the configuration and motion of the steam engine evolved from the atmospheric engine of Thomas Newcomen to James Watt's engine with separate condenser, from reciprocating motion (piston in cylinder) to rotating motion (turbine). This evolution is in accord with contemporary statements of evolutionary science made in biology, social sciences, and engineering. Improvements can be described as a procession of configurations that offer less dissipation in the engines, and correspondingly lower rates of irreversibility, or entropy generation (that is, a greater share of the useful energy received by the engine from the fuel is used for delivering work from the engine to the environment of the engine).

Second, on the brakes side of the Earth design, the evolution toward more W from the fixed Q_H means an evolution toward more and more Q_{diss} (work dissipated into heat). As the engines evolve toward more W, the brakes evolve toward dissipating more W, that is, toward more dissipation and higher rates of irreversibility or entropy generation. This aspect of constructal evolution agrees with statements heard in geophysics. Yet it is important to note that what geophysicists say (higher dissipation) is the complete opposite of what animal design and engineering scientists say (lower dissipation). The conflict between the two camps is real, but it is put to rest by the engines-and-brakes design of Figure 57, or its alternate shown in Figure 59. Both tendencies are manifestations of the single tendency expressed by the constructal law: All flow systems improve over time, so that we find the evolution of better-flowing engines (lower dissipation) and more effective brakes (higher dissipation).

In sum, the design of nature sketched in Figures 57 and 59 is an engine-and-brake system for using and destroying the useful energy (exergy) streaming to Earth from the sun. All the flow systems on Earth function as converters of useful energy (fuel or food) into mass moved. Nothing could be simpler. This view is a unifying theory because it encompasses the diverse domains

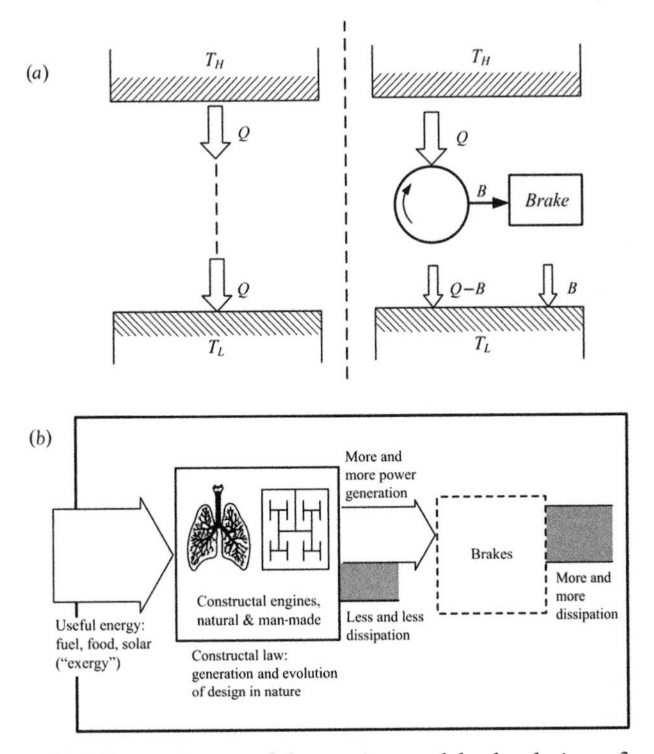

Figure 59. The evolution of the engine-and-brake design of nature. The constructal law governs how the flow system emerges and persists: by generating a flow architecture that distributes imperfections through the flow space and endows it with configuration. The "engine" part evolves in time toward generating more power (or less dissipation), and as a consequence, the "brake" part evolves toward more dissipation. (a) The original version of the engine-and-brake image of everything that moves on Earth. Q is the heat input to the engine, and B is the work output dissipated completely in the brake. (b) The engine-and-brake design of nature is represented by the flow of useful energy (exergy) into the Earth (the large rectangle), and by the partial destruction of this flow in the animate and inanimate engines (the larger square), followed by the complete destruction of the remaining useful energy stream in the interactions with the environment (the brakes shown in the smaller square). In time, all the flow systems exhibit the constructal-law tendency of generating designs, and this time the arrow points toward less dissipation in the engines and more dissipation in the brakes.

in which evolutionary phenomena are observed, recorded, and studied scientifically: animal design, river basins, turbulent flow, animal locomotion, speed records in athletics, technology evolution, and global design.

Once we know the global design of nature, we gain a new understanding of the emergence and evolution of life on Earth.

As the planet formed, there were many streams of heat (for example, lava flows, intense solar heating) that flowed straight from the warmer ground into the cooler ambient. As these heat streams rose, they mixed with the cooler, more dense fluids through the process of natural convection. In addition, some of these streams moved more quickly than others. When the flow was slow, it moved in a sheetlike motion, called laminar flow, because this was a good way to spread its momentum laterally. When the flow was fast enough, it transitioned to a turbulent flow because this was a more effective way to transfer that momentum laterally in the face of resistance.

How come? Because the tendency in nature is to equilibrate not only the hot with the cold but also the slow with the fast—equilibrium means uniformity in every respect. Thus, moving mass interacts with, *mixes* with, and *churns* all around it. When heat flows from high temperature to low temperature, and momentum flows from fast fluid to slow fluid, the result is not just the movement of mass but the mixing and churning of hot and cold, fast and slow. This flow moves in one direction—from the entity that possesses it (whether it's momentum, warmth, chemical species, knowledge, food, culture, etc.) to that which does not.

This is the familiar part of the story. What the constructal law teaches is that natural convection and turbulent flow are *designs*—the first designs—that emerged to facilitate the movement and mixing of currents (mass). They persist to this day because they still facilitate flow. In time, new structures emerged that could move currents faster and farther: rivers of lava, atmospheric and oceanic currents, and rain and rivers. Rainwater flowed even more easily by coalescing as rivulets and larger streams than when it was only seeping into the ground. Fast-flowing lava moved better through tree-shaped channels.

In addition, these living systems—these engines attached to brakes and driven by the sun—evolved in one direction in time, acquiring better and better designs for flowing, mixing, and churning. Remember, all the flows on Earth move actual currents of mass through actual channels. "Abstract" flows such as ideas, knowledge, science, information, technology, culture, and innovation are in fact facilitating the real flow of the human-and-machine species on the landscape. Easier flow, then, means *moving more mass (or weight) on Earth* by using the finite driving power derived from the sun. Because all these flows are tied to *geography*—because they all occur on the landscape of the Earth—this means that they morphed to spread their current over a larger area per unit of useful energy destroyed. This does not mean that every change was an improvement but that, broadly speaking, the changes that persisted were those that facilitated flow access.

In time, ever more complex designs emerged and evolved to facilitate and enhance global flow. The template for one of the most prevalent types of flow (from a point to an area or from an area to a point) is the river basins that cover the globe. All began as individual raindrops that coalesced when they could move more easily together. Over millions of years, everywhere around the planet, multiscale channels with a hierarchical, vascular design emerged to enhance flow both locally and globally. This design did not encompass just the black lines of the channels but the entire white area of the Earth where water seeped from the ground to those channels. It requires useful energy to create the channels of the river basin and to constantly move in relation to them, that is, to flow. The design that arose struck a balance between the time to move short and slow (seeping) and the time to move fast and long (in the channels). Governed by the constructal law, nature put the right pieces in the right places to facilitate more flow per unit of useful energy over the entire area.

This is also true of the Earth itself. This flow design—which encompasses all living systems on the planet—has also evolved with a single direction in time: to move more mass more easily. Its history is the story of the emergence and accumulation of these myriad evolving designs, these right channels put in the right places, to move more mass on a global scale. When we ask why any design exists—why we find a river basin, tree, or beetle—we should not consider it in isolation but see how it enhances global flow access. When we consider this evolutionary history, we should find that each new design has improved the flow that had existed. While it is true that new designs often absorb or replace those that had existed before, there are no winners and losers. All the engines are part of the single engine we call the global design, or nature, whose flows get better and better in time.

In big history, the inanimate designs that first emerged were eventually complemented by animate designs also driven by the sun, which further enhanced the mixing and churning of global flow. Even organisms that do not move because they are attached to other bodies (to rocks, for example, or animal skins) mix the medium that surrounds them because they draw in nutrients and oxygen and they expel products of metabolism. Animals that do not move set the environment in motion just as a sleeping person creates a plume of warm air that rises to the ceiling and mixes the room air.

As they took up space, moved, and excreted, the earliest single-cell organisms enhanced the churning process, albeit weakly. In 2009, researchers at Princeton and Northwestern confirmed this prediction of the constructal law when they reported that bacteria power microscopic "gears" when they are swimming. Churning means mixing of all kinds—of momentum, energy, and chemical species. Eventually, and inexorably, newer creatures evolved that were better at facilitating this mixing.

What is a fish but an eddy of water with its own motor inside? The eddy of water is the body of the fish moving. As it moves through the ocean, it displaces water. This churns the water, aiding its mixing and movement. The fish can also reach depths that the warm surface water cannot, thereby adding even more churning to the system. The fish is an eddy generator—many fish are paddle wheels mixing the water around their bodies. The swimmer in a lap pool is another generator of eddies, accomplishing the same thing in its own water environment. The lap pool with the swimmer in it is the brake with the engine, as in Figure 57.

Turning this example around, the big loop of ocean water driven between the equator and the poles is one Earth-size fish or swimmer.

The traditional view sees the movement of ocean currents and the swimming fish as separate phenomena. The constructal law teaches that fundamentally both have emerged as part of the same global design, working together to enhance flow and mixing on Earth.

This insight allows us to see the whole design of evolution. Just as we predict that all the rain falling on the ground should configure itself into rivulets, streams, and, eventually, main channels, biological life should evolve to make the whole Earth flow more easily.

All animals, regardless of their habitat, mix air and water more efficiently in the presence of an existing flow structure. It sounds crude, but this is what biological flow systems accomplish and why their legacy is the same as that of the rivers and the winds. Animals move mass from here to there. Animals in tandem with air, water, sand, and dust move more mass than the inanimate currents did before animate systems emerged.

This does not negate the idea that biological life sprang from the primordial soup, because the chemical conditions were right for that. Things flow on Mars, but Mars does not support biological life because the conditions aren't right there. Perhaps not right yet or not right anymore. The constructal law does not predict that a biosphere should exist on Earth and not on another planet. Similarly, it does not predict that there should be river basins. Environment matters. What it does predict is that if the conditions exist for a biosphere or river basin to arise, then they should acquire configurations that facilitate movement and mixing.

Thus the constructal law reveals, for the first time, a predictable through-line for the history of Earth. At each stage of this story—the rise of the lithosphere, atmosphere, hydrosphere, and biosphere—nature evolved to facilitate the movement of more mass on Earth. Instead of seeing living things only as isolated forms trying to find a niche and ensure their own survival, the constructal law teaches us that they have evolved as manifestations of the tendency of all things to enhance global mixing and churning in accordance with the laws of thermodynamics and the constructal law. The animate designs of the biosphere are newer and they emerged because they complement—they enhance—the mixing performance of the inanimate designs.

Because animate phenomena are not a break from, but a continuum of, the evolutionary process that began with inanimate phenomena, we predict similar patterns in both. The hierarchical, vascular design that characterizes suddenly emerging lightning bolts and slowly evolving river basins is what we also find in our circulatory and respiratory systems. Scaling laws that determine the relationship between all the streams in all the river basins and all the blood vessels in our bodies also predict the relationship between mass and speed that we find in all animals. And we find a predictable distribution of channels not only in a lava flow but also in the rankings of trees in the forest, and of cities, universities, language, and other phenomena that spread a current across a volume or an area.

The argument—until now—has been about whether measurable improvement can be found in the evolution of biological organisms. By this, I mean something more than the idea that increasing complexity has marked evolution. Have animals indeed become better?

The constructal law provides a resounding yes to this question.

At every stage of biological evolution, the evolving characteristics that "stuck" were those that measurably enhanced movement of mass on Earth. All of them. As we saw in chapter 3, land animals require less work to cover a certain distance than the sea creatures that came before them. Similarly, insects and birds need less work to cover the same distance as land animals of the same weight. In addition, bigger animals expend less useful energy (work, exergy) per kilogram of body mass moved than smaller ones. Elephants need to eat far more than dogs to generate the energy required to cover distance, but kilo for kilo they need far less.

The constructal law transforms our understanding of evolution in other ways. Was it inevitable that animals should fill the sea, land, and air? Yes, because evolution toward greater access means that flow systems should evolve to penetrate, mix, and churn larger areas and volumes. To take the simplest example, we can predict the number of channels of lava that a volcano should form if we know the area that lava will cover—the bigger the area, the greater the number of channels and the greater number of branching levels, that is, the greater the complexity. And because we know the area size, we can predict the complexity and the fact that complexity must be finite, i.e., modest.

This same principle enables us to predict why we should find so much diversity in biological organisms. Some of the apparent diversity—the multiple scales and numbers—is organized, and the name for that is hierarchy. Just as river basins have multiscale channels and the forest floor is covered by vegetation of varying sizes, biological organisms of various sizes emerge to spread mass over the planet. As we discussed in chapter 6, flow systems with hierarchical shape and structure have few large components and many small ones because this is a good way to spread a current over an area or volume. So it is with the broad pattern of biological life: The smaller organisms are, the greater their population. Researchers estimate that there are perhaps ten times as many bacteria as human cells in our bodies. The British entomologist

C. B. Williams once estimated that there are perhaps one million trillion insects alive on the planet at any one time.

This hierarchical design has often been described in terms of the food chain—generally speaking, big animals do feed on smaller ones that prey on ever-smaller ones. This is true as far as it goes. The problem is that it casts animals in the context of codependent competitors vying against one another for survival. There is a balance being struck in this relationship, but it is not one that should be understood in the context of a particular species. Instead, we find diversity in the size and distribution of animals because this is a good design for covering an area with the movement of animal mass—all kinds of animal mass, including all kinds of trucks carrying our loads. Big animals and big trucks may be more efficient than smaller ones in terms of the expenditure of useful energy, and they also mix and churn more of the environment. But smaller animals can penetrate spaces that larger ones cannot. This is the design that sustains all ecosystems. It takes all kinds.

This raises the question: Does the story of evolution from inanimate flows (alone) to inanimate and animate flows (together) end with the emergence of the biosphere? The answer is no, for two reasons. First, throughout this book we have noted that all flow systems will morph if "given freedom." This is a necessary qualification because, in the global design, the brakes are as essential and as free to morph as the engines. What the constructal law reveals is that freedom is natural. Freedom is what allows flow systems to configure and reconfigure themselves. It is what allows them to "get design" and get better. Without freedom there would be no design and no evolution.

So it goes with human beings, who are a piece of this evolutionary tapestry of designs (Figure 60). Like all that came before us, we, too, are flow systems for mixing and churning the Earth. Our great achievements—hunting, agriculture, religion, science, medicine, government, art, commerce, etc.—all reflect the tendency to generate configurations to move our mass on Earth,

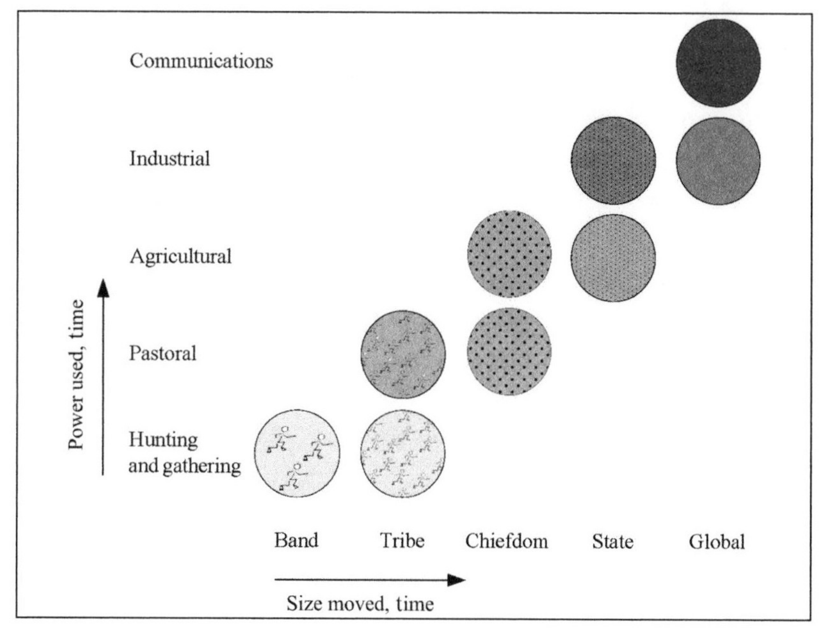

Figure 60. The evolving designs of the human-and-machine species on Earth are known by more familiar names. In time, the territory swept by organized members of the species increases from the band, tribe, and chiefdom to the state and the global design today. At the same time, the production and use of power (Figure 58) increase along with other features that characterize the sizes of the streams driven by the consumed power: subsistence, standard of living, rate of fuel consumption, traffic, gross domestic product, wealth, affluence, advancement, etc.

and to reshape the Earth. Indeed, the recorded history of mankind chronicles the invention of better and better designs for this.

The rise of civilization, the spread of individual liberty and empowerment, the emergence of technology, the wider spread of goods and ideas, this is the story of humanity: the creation of new and better systems that flow across the Earth's crust. Since the Industrial Revolution we have been witnessing the emergence of an additional design that enhances performance of the atmosphere, hydrosphere, and biosphere. The new sphere is the global flow of *human-and-machine species*. Each of us is much more than a naked body. Because the *homo* is *sapiens*, life is becoming easier during our lifetime. From the mirror to the bookshelf and the street, we have enlarged ourselves with knowledge, which means everything: machines, shelter, food, and links all over the globe. We are flowing internally, in our organs. We also flow externally, through every move and sound, from the planted fields to the school yard, from the horse to the airplane, from the telephone to the Internet.

If this is a book about design in nature, then why mix nature with engineering? The reason is that we are an integral part of nature. The engineered or the "artificial" is the realm of the objects that exist only because we exist. They do so as extensions of our own bodies and actions. Because we are natural (we happen), our extensions are natural, not unnatural. Thanks to engineering in all its forms (technology, medicine, business, education, communications, government), each of us is much larger, stronger, and faster than our naked bodies. Each of us is as big as the global sphere.

The human-and-machine species we have become is evolving visibly every day (for example, Figure 58). Its evolution is even more obvious if we reexamine history and geography, including the history of science, technology, and individual liberty. Life is monumentally easier for those who possess knowledge.

Human evolution, yes, but in what direction? In the same direction as the evolution of all other flow configurations: toward designs that flow more easily as a whole. All the mass that moves with us and because of us (people, goods, information) is flowing more easily with progressively greater access all over the map: from human migrations in history to globalization and free trade today, from global electrification in the twentieth century to mass air travel and communication in the twenty-first century; from slaves and serfs tied to the ground to free individuals and individuals on vehicles.

More mass moved to greater distances for every unit of fuel (food) and effort. This is what the evolution of the human-and-machine design is achieving. It is also what the evolution of animal design and the entire biosphere is achieving. If the biosphere were to come to rest completely tomorrow, so dead that *nothing moves*, the legacy of all the biosystems on Earth is this: Mass moved and the Earth's crust mixed. It is the legacy of the rivers that dry up and of civilizations that vanish. All dead flows are pyramids.

People often chafe at restrictions—we want to be free! It may seem grim and confining to see ourselves as just another flow system in nature destined to find better and better ways to move mass. The silver lining is that evolutionary history has aligned our destiny with our desires, and this is no coincidence. Our impulses, thoughts, and actions are geared toward movement and flow. The basic instincts of humanity—safety, nourishment, health, mating, longevity—are expressions of this constructal urge. So, too, are our greatest creations, including science, technology, government, economics, and the arts. All are flow systems that have emerged and continue to evolve in order to facilitate the movement of mass on Earth. All are evolving in time to increase access for their currents. All are part of the tapestry of mating and morphing flow designs that cover the Earth.

The discovery of the constructal law does not change or diminish our aims, accomplishments, or pleasures. It finally explains what drives our history—far better than sexual reproduction (again, what DNA is responsible for the lungs of the Earth, the river basins, the vegetation, and the lightning trees?). From the hunter-gatherer tribes to industrial farms, from the wheel to airplanes, and from smoke signals to the Internet, we have developed, then improved upon, designs to enhance our movement. The constructal law unites that history with all around us. Suddenly, we make sense in the grand scheme; we are not apart from nature but another manifestation of its oneness.

Most rewarding in this holistic view of the design of movement on Earth is how simple, universal, and familiar the direction is. Over time, inanimate and animate designs have evolved to use energy more efficiently to move more mass on Earth. Seen constructally, our planet is an enormous river basin driven by the sun with a hierarchical distribution of evolving multiscale animate and inanimate channels, whose size and distribution are balanced to facilitate flow. No law of thermodynamics calls for this design and design change. Yet design evolution happens because it is an integral part of physics. This is the part of physics covered by the constructal law.

As the constructal law describes the past, it also allows us to predict the future. What's ahead? The short answer should be obvious by now: a multitude of flow designs that move more mass better—cheaper, farther, faster. On the world stage, you can place solid bets that the entire globe will continue spreading the rule of law, free trade, human rights, globalization, and all the other design features that guarantee more movement for us and our stuff. Sure, there are obstacles; dictators will not like this prediction. But the nature of the flow system composed of the huge numbers of individuals makes their reigns short-lived in the grand scheme—because of physics.

In addition, the human-and-machine species we have become will continue to evolve before our eyes. Relatively new technologies—such as cell phones and handheld computers that have increased and improved the flow of information, people, and goods across the globe—will be complemented by new inventions that will allow our currents to sweep the globe more easily, more cheaply. The modern-day Thoreaus who lament the "dehumanizing" effects of these natural developments are not just on the wrong side of history but also on the wrong side of physics.

So, too, are those who demand that the world's population reduce its use of energy. Power technology will continue to evolve toward greater efficiencies and more power produced and used, not less. The pursuit of higher efficiency will not lead to less fuel consumption. The evidence for this is massive. The direction has always been one-way: more power for more individuals over larger territories, and more power used by every individual.

Throughout human history, when one source of power proved

insufficient, a new one was *added*—first people, then people with animals, then medieval contributions from windmills and water wheels. The old sources of power were not abandoned. The big change came with the development of heat engines. This spurred two revolutions: the industrialization and electrification of the globe and the empowering of science with a great and entirely new discipline that came from engineering: thermodynamics. Steam engines were joined by many kinds of power plants in the late 1800s and the 1900s: steam turbine, gas turbine, hydroelectric, nuclear, solar, eolian, ocean thermal, ocean waves, etc. Fuels, too, have become more diverse, from coal and waterfalls to petroleum, nuclear fuel, solar heating, geothermal, and wind.

Our movement is proportional to the amount of fuel that we burn. It represents everything that characterizes us as live beings: transportation, trade, economics, business, and communications. This is why the GDP values of all countries and political regions of the globe are proportional to their respective amounts of fuel usage. Movement is wealth, and wealth is physics. "Getting lucky" is a manifestation of design.

Like the movement of everything else, our movement is a living shell that thrives on the whole globe. This organism has a heart with two chambers, Europe and North America; several vital organs in the Far East; and a vascular tissue that covers the entire inhabited globe (see Figure 42, p. 179). Scientific projections show that in 2050 the organism will have the same structure, but all the streams will be thicker in comparison with what they are today. By understanding that energy usage is not simply a political or social problem but a natural phenomenon that governs human actions, we see the issue in truly global terms. It teaches us that power flows with predictable design, which means configuration, pattern, rhythm, and geography. The design evolves and grows. It flows in an evolutionary manner, in time, as thicker and more efficient streams that serve (and liberate) humanity over greater territories. Thus, the surge of interest in global energy sustainability, green solutions, and wind power is not a new mode of thinking but just the latest manifestation of the tendency that governs flow on Earth: the evolution of better designs to move more, not less, mass on Earth.

Finally, because the constructal law illuminates design in nature and the direction of evolution, I predict that people—more and more, over larger and larger areas—will use this powerful discovery to generate fresh insights about the things that matter to them. Civilization with all its constructs (science, religion, language, writing, and so on) is the never-ending physics of evolving flow configurations from the movement of mass, energy, and knowledge to the world migration of people to whom ideas occur.

Good ideas travel and keep on traveling. Better-flowing configurations replace existing configurations. That is life. That is our history. That is the future.

ACKNOWLEDGMENTS

During the past sixteen years, I have had many collaborators in developing the constructal law: Sylvie Lorente, Heitor Reis, António Miguel, Stephen Périn, Gilbert Merkx, Cesare Biserni, Luiz Rocha, Shigeo Kimura, Giulio Lorenzini, Tanmay Basak, James Marden, Jordan Charles, Edward Jones, Erdal Cetkin, Yong Sung Kim, Jaedal Lee, Kuan-Min Wang, Atit Koonsrisuk, Sunwoo Kim, and many more. This book is about our most recent work together and a warm reminder that science is good and fun when interesting ideas bring interesting minds together.

The text and figures were organized and improved by Deborah Fraze. New figures were drawn by Erdal Cetkin and Kuan-Min Wang.

Thanks are also due to Janine Steel Zane, Suzanne Pederson, Phillip Manning, Thomas G. Hentoff, E. Kemp Reece Jr., Dr. Kitty M. Steel, and Lewis Steel.

My coauthor and I owe a special debt of gratitude to our agent, Tina Bennett. She heard the click early on, and her insightful guidance every step of the way has helped this project flow better and better.

Finally, we owe special thanks to the team at Doubleday, beginning with our editor, Melissa Danaczko, whose sharp eye and boundless curiosity improved this book in ways large and small.

NOTES

Introduction

7 **published hundreds of articles:** Bejan, A., and Lorente, S., 2004, The constructal law and the thermodynamics of flow systems with configuration, *International Journal of Heat and Mass Transfer*, vol. 47, pp. 3203–3214; Bejan, A., and Lorente, S., 2005, *La Loi Constructale*, L'Harmattan, Paris; Bejan, A., and Lorente, S., 2006, Constructal theory of generation of configuration in nature and engineering, *Journal of Applied Physics*, vol. 100, article 041301; Bejan, A., and Lorente, S., 2010, The constructal law of design and evolution in nature, *Philosophical Transactions of the Royal Society B*, vol. 365, pp. 1335–1347; Bejan, A., and Lorente, S., 2011, The constructal law and the evolution of design in nature, *Physics of Life Reviews*, vol. 8; Biserni, C., Rocha, L. A. O., Stanescu, G., and Lorenzini, E., 2007, Constructal H-shaped cavities according to Bejan's theory, *International Journal of Heat and Mass Transfer*, vol. 50, pp. 2132–2138; Lewins, J., 2003, Bejan's constructal theory of equal potential distribution, *International Journal of Heat and Mass Transfer*, vol. 46, pp. 1541–1543; Lorenzini, G., 2006, Constructal design of Y-shaped assembly of fins, *International Journal of Heat and Mass Transfer*, vol. 49, pp. 4552–4557; Miguel, A. F., 2006, Constructal pattern formation in stony corals, bacterial colonies and plant roots under different hydrodynamic conditions, *Journal of Theoretical Biology*, vol. 242, pp. 954–961; Miguel, A. F., 2010, Natural flow systems: acquiring their constructal morphology, *International Journal of Design & Nature and Ecodynamics*,

vol. 5, no. 3, pp. 230–241; Poirier, H., 2003, Une théorie explique l'intelligence de la nature, *Science & Vie*, no. 1034, pp. 44–63; Quéré, S., 2010, Constructal theory of plate tectonics, *International Journal of Design & Nature and Ecodynamics,* vol. 5, no. 3, pp. 242–253; Raja, V. A. P., Basak, T., and Das, S. K., 2008, Thermal performance of a multi-block heat exchanger designed on the basis of Bejan's constructal theory, *International Journal of Heat and Mass Transfer,* vol. 51, pp. 3582–3594; Reis, A. H., and Gama, C., 2010, Sand size versus beachface slope—An explanation based on the Constructal Law, *Geomorphology,* vol. 114, p. 276; Rocha, L. A. O., Lorenzini, E., and Biserni, C., 2005, Geometric optimization of shapes on the basis of Bejan's Constructal theory, *International Communications in Heat and Mass Transfer,* vol. 32, pp. 1281–1288; www.constructal.org.

7 **My own books for specialists:** This book is my first effort to share the major findings of my work with a broad audience; I encourage anyone who wants to access my scholarly publications to visit www.ISIhighlycited.com.

10 **"maximized":** Although they are used widely, terms such as "maximized," "minimized," and "optimized" have no place in discussion of design in nature. The tendency in nature is not to generate the "most efficient" or "best" design but one that works and gets better over time. The tendency of nature is *evolutionary* design generation.

11 **Before the constructal law:** Bejan, A., and Gobin, D., 2006, Constructal theory of droplet impact geometry, *International Journal of Heat and Mass Transfer,* vol. 49, pp. 2412–2419.

15 **Three colleagues and I found:** Bejan, A., Lorente, S., Miguel, A. F., and Reis, A. H., 2006a, Constructal theory of distribution of river sizes, Section 13.5 in Bejan (2006).

16 **globalization:** Bejan, A., and Lorente, S., 2011, The constructal law and the evolution of design in nature, *Physics of Life Reviews,* vol. 8.

16 **warfare:** Weinerth, G., 2010, The constructal analysis of warfare, *International Journal of Design & Nature and Ecodynamics,* vol. 5, no. 3, pp. 268–276.

16 **mortality:** Manton, K. G., Land, K. C., and Stallard, E., 2007, Human aging and mortality, chapter 10 in Bejan and Merkx (2007).

18 **Many other scientists have offered:** Glazier, D. G., 2005, Beyond the ¾-power law: Variation in the intra- and interspecific scaling of metabolic rate in animals, *Biological Reviews,* vol. 80, pp. 611–662; Hoppeler, H., and Weibel, E. R., 2005, Scaling functions to body size, *Journal of Experimental Biology,* vol. 208 (9), special issue;

Mandelbrot, B. B., 1982, *The Fractal Geometry of Nature*, New York: Freeman; Schmidt-Nielsen, K., 1984, *Scaling (Why Is Animal Size So Important)*, Cambridge, UK: Cambridge University Press; Vogel, S., 1988, *Life's Devices*, Princeton, NJ: Princeton University Press; Weibel, E. R., 2000, *Symmorphosis: On Form and Function in Shaping Life*, Cambridge, MA: Harvard University Press.

18 **minimum, maximum, optimum:** For reviews, see Bejan and Lorente 2010, 2011; Bejan, A., 1996, *Entropy Generation Minimization*, Boca Raton, FL: CRC Press; Bejan, A., 1997, *Advanced Engineering Thermodynamics,* 2nd ed., New York: Wiley, chapter 13.

19 **Thermodynamics rests on two laws:** Bejan, A., 1982, *Entropy Generation Through Heat and Fluid Flow*, New York: Wiley, p. 35.

20 **"pernicious tendency":** Turner, J. Scott, *The Tinkerer's Accomplice: How Design Emerges from Life Itself,* Cambridge, MA: Harvard University Press, 2007.

20 **"The idea that there is 'design' in nature":** "There Is 'Design' in Nature, Brown Biologist Argues at AAAS," Brown University press release, February 17, 2008, http://news.brown.edu/pressreleases/2008/02/aaasmiller.

21 **"how [complicated things] came into existence":** Dawkins, Richard, *The Blind Watchmaker: Why the Evidence of Evolution Reveals a Universe Without Design*, New York: W. W. Norton, 1986.

22 **"A lot has been written about natural selection":** Bejan, A., 1997, Constructal-theory network of conducting paths for cooling a heat generating volume, *International Journal of Heat and Mass Transfer* 40 (published on November 1, 1996), 799–816.

24 **"'living' is motion":** Faulkner, William, *The Mansion,* New York: Vintage International Edition, 2011.

24 **"Dwell as near as possible":** Thoreau, Henry David, *The Journal of Henry D. Thoreau: In Fourteen Volumes Bound as Two, Volumes I–VII,* Toronto: Dover Publications, 1962.

24 **"The fundamental principle of human action":** George, Henry, *Progress and Poverty,* New York: D. Appleton and Company, 1886.

CHAPTER 1
The Birth of Flow

33 **into the surrounding tissues:** Reis, A. H., Miguel, A. F., and Aydin, M., 2004, Constructal theory of flow architecture of the lungs, *Medical Physics*, vol. 31, pp. 1135–1140.

40 **"the sum total of things":** Letter to Herodotus.

44 ***Freedom is good for design*:** Bejan, A., and Lorente, S., 2010, The constructal law of design and evolution in nature, *Philosophical Transactions of the Royal Society B*, vol. 365, pp. 1335–1347.

45 **the second law of thermodynamics:** When reading about the second law, most of us think the next word that we will see is "entropy." Not in this book, although an introduction to this language is useful. The reason is that entropy is not necessary in order to express the natural tendency summarized by the second law. Entropy is the system property that Clausius had to define (by relying on the second law) in order to state the second law analytically, as a mathematical inequality. Much of the confusion that surrounds the second law today stems from entropy mathematics, which sounds impressive but adds nothing to the physics. Clausius and his contemporary Lord Kelvin preferred to state the second law in words, not math, as did the giants of the generations that followed (for example, Max Planck and Henri Poincaré). The second law must not be confused with mathematical formulas and properties (entropy) defined later in order to facilitate the practical use of the second law.

49 **Reynolds number:** The Reynolds number (Re) is a dimensionless number calculated by multiplying the velocity of the flow (V) with a length scale of the flow, and dividing this product by the kinematic viscosity of the fluid (ν), $\mathrm{Re} = V \times \mathrm{Length}/\nu$. In the constructal prediction of turbulence, V is the *longitudinal* velocity of the flow, and the length scale is the *transversal* dimension (the thickness) of the flow, D. In other words, $\mathrm{Re} = VD/\nu$.

50 **reaches this threshold:** Bejan, A., 1982, *Entropy Generation Through Heat and Fluid Flow*, New York: Wiley, p. 35.

51 **"Often a very obvious thing may lie unnoticed":** Steinbeck, John, *The Log from the Sea of Cortez,* New York: Penguin, 1986.

51 **If nothing interacts with the box:** An isolated system should not be confused with a closed system (see Figure 8). A system is isolated if it has absolutely no interactions with its environment, that is, no mass flows, no heat transfer, and no work transfer. A closed system does not have mass flow interactions with its environment, but it may have heat and work interactions. An isolated system is closed, but a closed system is not necessarily isolated.

CHAPTER 2
The Birth of Design

54 **"Design (drawing) . . . is the root of all sciences":** de Hollanda, Francisco. *Four Dialogues on Painting*, trans. Aubrey F. G. Bell. London: Oxford University Press, 1928.

61 **our morning commute:** Lorente, S., and Bejan, A., 2010, Few large and many small: hierarchy in movement on Earth, *International Journal of Design & Nature and Ecodynamics*, vol. 5, no. 3, pp. 254–267.

62 **not fully appreciated before the constructal law:** Fractal geometry, for example, focuses only on the channels (which, famously in fractal geometry, fill the space "incompletely"). Fractal geometry does not focus on the design of the interstices. It cannot possibly illuminate design in nature because it ignores half the design: the white between the black lines. In the real world, flow designs emerge to serve an entire area or volume.

67 **In the second experiment:** Chen, J.-D., Radial viscous fingering patterns, *Exp. Fluids*, vol. 5, pp. 363–371.

69 **sand in a laboratory:** Bejan, A., 1997, *Advanced Engineering Thermodynamics,* 2nd ed., New York: Wiley, chapter 13.

70 **This is evolution, reproduced in the laboratory:** Parker, R. S., "Experimental Study of Drainage Basin Evolution and Its Hydrologic Implications," Colorado State University, Hydrology Paper 90, Fort Collins, CO, 1977.

75 **He hatched chickens:** This was his protest and resistance. The communist regime insisted that all the means of production must belong to the state. Through his actions, my father questioned authority by demonstrating the absurdity and inhumanity of the prevailing policy. For me, this was the lesson for how to be a scientist. The most effective way to question prevailing dogma is to demonstrate the better idea, naked on the table, without fear of punishment from the establishment. Punishment will come, and when it does it is a compliment.

CHAPTER 3
Animals on the Move

81 ***Larger animals are faster*:** It is no coincidence that larger rivers and trucks are also faster than smaller ones. (Lorente, S. and Bejan, A.,

2010, Few large and many small: hierarchy in movement on Earth, *International Journal of Design & Nature and Ecodynamics*, vol. 5, no. 3, pp. 254–267.)

86 **This work is equal to the drag force:** The drag force has been studied extensively, and its scale is $F \sim \rho_m V^2 L_b{}^2$, where V is the air speed, and ρ_m is the density of the medium (for air, $\rho_m \sim 1$ kg/m^3).

86 **the timescale of one cycle:** The timescale, t, is dictated by the time needed to fall vertically to the distance L_b (recall that this fall was the first step in the cycle scenario). The timescale of free fall is $(L_b/g)^{1/2}$, the Galilean timescale of free fall.

88 **for my 2000 book:** Bejan, A., 2000, *Shape and Structure, from Engineering to Nature*, Cambridge, UK: Cambridge University Press.

89 **closer to a cycloid:** The cycloid is the cyclical "hopping" curve traced by a point on the rim of the wheel as it rolls on the ground. I use this word because this is what the animal is: the point on its own "animal wheel" that rolls horizontally. (Bejan, A., 2010, The constructal-law origin of the wheel, size, and skeleton in animal design, *American Journal of Physics*, vol. 78, no. 7, pp. 692–699.)

95 **when plotted on a log-log graph:** A function that has the form $y=cx^k$, where c and k are constants, can also be written as $\log y = \log c + k \log x$, which is a linear relation between $\log y$ and $\log x$. Consequently, if the original function $y(x)$ is plotted on a graph with $\log y$ versus $\log x$, the function will appear as a straight line with the slope k.

96 **in bulk increases with size:** Lorente, S., and Bejan, A., 2010, Few large and many small: hierarchy in movement on Earth, *International Journal of Design & Nature and Ecodynamics*, vol. 5, no. 3, pp. 254–267.

CHAPTER 4
Witnessing Evolution

103 **Since 2003:** Bejan, A., and Lorente, S., 2008, *Design with Constructal Theory*, Hoboken, NJ.

104 **His first set of data:** Charles, J. D., and Bejan, A., 2009, The evolution of speed, size and shape in modern athletics, *Journal of Experimental Biology*, vol. 212, pp. 2419–2425.

107 **"the closest thing to a grand unified theory":** Futterman, Matthew, "Behind the NFL's Touchdown Binge," *The Wall Street Journal*, September 9, 2009.

107 **Charles and I concluded:** Charles, J. D., and Bejan, A., 2009, The

evolution of speed, size, and shape in modern athletics, *Journal of Experimental Biology,* vol. 212, pp. 2419–2425.

108 **This puzzle was proposed:** Bejan, A., Jones, E. C., and Charles, J. D., 2010, The evolution of speed in athletics: why the fastest runners are black and swimmers white, *International Journal of Design & Nature and Ecodynamics*, vol. 5, no. 3, pp. 199–211.

116 **The natural emergence of the wheel:** Bejan, A., 2010, The constructal-law origin of the wheel, size, and skeleton in animal design, *American Journal of Physics*, vol. 78, no. 7, pp. 692–699.

120 **When a force is applied suddenly:** Here we find another example of how predictable patterns emerge and evolve in a flash to facilitate flow.

CHAPTER 5
Seeing Beyond the Trees and the Forest

126 **"I went to the woods":** Thoreau, Henry David, 2004, *Walden: A Fully Annotated Text*, ed. Jeffrey S. Cramer, New Haven, CT, and London: Yale University Press, p. 88.

127 **"all religions, arts, and science":** Einstein, Albert, 1976, *Out of My Later Years,* New York: Citadel Press, p. 9.

131 **predicts the design of a tree:** Bejan, A., Lorente, S., and Lee, J., 2008, Unifying constructal theory of tree roots, canopies and forests, *Journal of Theoretical Biology*, vol. 254, pp. 529–540.

132 **resistivity:** Resistivity must not be confused with resistance. Resistance is a phenomenon that a current encounters. Resistivity is a property of the material itself. The resistance of an electrical conductor is proportional to the length of the conductor times the resistivity of the conductor material. Resistance is a property of the flow configuration, the design. The fluid flow resistance of a capillary tube is proportional to the length (L) of the tube multiplied by the viscosity of the fluid divided by the tube diameter (D) raised to the power of 4. In this fluid flow, the viscosity is the material property that plays the role of resistivity, whereas the flow resistance is a property of the flow configuration because it is also proportional to L/D^4. The configurations of nature are sharply visible because the high-resistivity portion flows hand in glove with the low-resistivity portions. Visibility and contrast happen because high resistivity is not the same as low resistivity. Water seeping through riverbanks cannot be

confused with water flowing down a river. Yet the constructal law reveals a subtlety that underpins the whole design: The *resistances* are the same, even though the two portions (seepage across wet riverbanks versus river channel flow) are famously different because, among other things, their resistivities are highly dissimilar.

132 **least resistance as a whole:** Bejan, A., Lorente, S., and Lee, J., 2008, Unifying constructal theory of tree roots, canopies and forests, *Journal of Theoretical Biology*, vol. 254, pp. 529–540.

134 ***stresses flow* through an object:** Bejan, A., Lorente, S., and Lee, J., 2008, Unifying constructal theory of tree roots, canopies and forests, *Journal of Theoretical Biology*, vol. 254, pp. 529–540; Lorente, S., Lee, J., and Bejan, A., 2010, The "flow of stresses" concept: the analogy between mechanical strength and heat convection, *International Journal of Heat and Mass Transfer*, vol. 53, pp. 2963–2968.

139 **harmony born of beautiful balance:** Bejan, A., 2009b, The Golden Ratio predicted: vision, cognition and locomotion as a single design in nature, *International Journal of Design & Nature and Ecodynamics*, vol. 4, no. 2, pp. 97–104.

CHAPTER 6
Why Hierarchy Reigns

151 **As my Duke University colleague:** "Language is a flow system," Bejan, A., and Merkx, G. W., eds., 2007, *Constructal Theory of Social Dynamics*, New York: Springer; Lorente, S., 2007, Tree flow networks in urban design, chapter 3 in Bejan and Merkx (2007); Morroni, F., 2007, Constructal approach to company sustainability, chapter 14 in Bejan and Merkx (2007); Périn, S., 2007, The constructal nature of the air traffic system, chapter 6 in Bejan and Merkx (2007); Staddon, J. E. R., 2007, Is animal learning optimal?, chapter 8 in Bejan and Merkx (2007); Tiryakian, E. A., 2007, Sociological theory, constructal theory, and globalization, chapter 7 in Bejan and Merkx (2007).

161 **What we find is an evolving architecture of channels:** This contradicts the popular claim made by proponents of the "fractal geometry of nature." They assert that tree design is fractal because if we zoom in on a subvolume of the big system, we rediscover the flow architecture of the big system. This is not true, because if we zoom in on a tree in the garden, we do not see a tree but the

empty space between the two smallest branches. No tree-shaped flow of nature is a fractal object. The geometry of nature is not fractal.

164 **"A new scientific truth":** Planck, Max, 1949, *Scientific Autobiography and Other Papers*, trans. Frank Gaynor, New York: Philosophical Society.

166 **Recall our discussion:** Bejan, A., Lorente, S., Miguel, A. F., and Reis, A. H., 2006a, Constructal theory of distribution of river sizes, Section 13.5 in Bejan (2006).

167 **This dovetailed with the findings:** Bejan, A., Lorente, S., Miguel, A. F., and Reis, A. H., 2006b, Constructal theory of distribution of city sizes, Section 13.4 in Bejan (2006).

174 **if we reran Stephen Jay Gould's:** Bejan, A., and Marden, J. H., 2006, Unifying constructal theory of scale effects in running, swimming and flying, *Journal of Experimental Biology*, vol. 209, pp. 238–248.

CHAPTER 7
The Fast and Long Meets the Slow and Short

175 **Why does it work?:** Bejan, A., 2006, *Advanced Engineering Thermodynamics*, 3rd ed., Hoboken, NJ: Wiley.

177 **The time to move fast and long:** Bejan, A., 2000, *Shape and Structure, from Engineering to Nature*, Cambridge, UK: Cambridge University Press; Bejan, A., and Lorente, S., 2006, Constructal theory of generation of configuration in nature and engineering, *Journal of Applied Physics*, vol. 100, article 041301.

182 **cooling systems for electronics:** Bejan, A., 1997, *Advanced Engineering Thermodynamics,* 2nd ed., New York: Wiley, chapter 13.

185 **The same rhythmic design:** Bejan, A., 1997, *Advanced Engineering Thermodynamics,* 2nd ed., New York: Wiley, chapter 13; Bejan, A., 2000, *Shape and Structure, from Engineering to Nature*, Cambridge, UK: Cambridge University Press.

185 **the Pyramids in Egypt:** Bejan, A., and Périn, S., 2006, Constructal theory of Egyptian pyramids and flow fossils in general, Section 13.6 in Bejan, 2006.

187 **With these parameters in mind:** Bejan, A., and Lorente, S., 2001, Thermodynamic optimization of flow geometry in mechanical and civil engineering, *Journal of Non-Equilibrium Thermodynamics*, vol. 26, pp. 305–354.

CHAPTER 8
The Design of Academia

200 **_What_ flows through a design:** Bejan, A., 1997, *Advanced Engineering Thermodynamics,* 2nd ed., New York: Wiley, chapter 13; Bejan, A., 2000, *Shape and Structure, from Engineering to Nature,* Cambridge, UK: Cambridge University Press.

201 **Each new release:** Bejan, A., 2007, Why university rankings do not change: education as a natural hierarchical flow architecture, *International Journal of Design & Nature and Ecodynamics*, vol. 2, no. 4, pp. 319–327.

202 **the absence of change stands out:** We find a similar rigidity of rankings in the World University Rankings compiled annually since 2004 by the *Times* of London. Higher education is a global flow.

213 **fans of college basketball:** Bejan, A., and Haynsworth, P., 2011, The natural design of hierarchy: basketball versus academia, *International Journal of Design & Nature and Ecodynamics*, vol. 6.

218 **dark networks in my paper:** Bejan, A., 2009a, Two hierarchies in science: the free flow of ideas and the academy, *International Journal of Design & Nature and Ecodynamics*, vol. 4, no. 4, pp. 386–394.

219 **today's research landscape:** Bejan, A., 2008, Constructal self-organization of research: empire building versus the individual investigator, *International Journal of Design & Nature and Ecodynamics*, vol. 3, no. 3, pp. 177–189.

220 **the few large and many small:** Lorente, S., and Bejan, A., 2010, Few large and many small: hierarchy in movement on Earth, *International Journal of Design & Nature and Ecodynamics*, vol. 5, no. 3, pp. 254–267.

CHAPTER 9
The Golden Ratio, Vision, Cognition, and Culture

222 **Countless generations:** Bejan, A., 2009b, The Golden Ratio predicted: vision, cognition and locomotion as a single design in nature, *International Journal of Design & Nature and Ecodynamics*, vol. 4, no. 2, pp. 97–104.

223 **Seen constructally, shapes that resemble the golden ratio:** The golden ratio is a mathematical object that mathematicians have considered through the ages, finding all kinds of properties in the

"golden number"—mathematical correspondences with Fibonacci numbers, etc. Here we don't address these mathematical connections; our work deals with physical phenomena in our line of sight. The mathematical games played with the golden number remain unaffected by our discovery, except that now the physics phenomenon (which intrigued the mathematicians in the first place) has its foundation in the same principle of physics that underpins the whole animal design: the constructal law.

225 **The evolution of writing, toward simplicity and universality:** Language is a flow system for spreading information, a phenomenon that my student Cyrus Amoozegar explored in "Constructal Theory of Written Language." In written languages, the constructs (pictographs, characters, letters, and symbols) are the channels through which those currents of information move. The evolution of written languages began with the pictographs that include the cave paintings from the prehistoric period. They depict various images of animals and humans and their meaning lies directly in what is painted. While this method avoided confusion—a rose is a rose is a rose—it was inefficient because every concept required its own drawing. (Amoozegar, C., 2007, Constructal theory of written language, chapter 16 in Bejan and Merkx [2007]; Bejan, A., and Merkx, G. W., eds., 2007, *Constructal Theory of Social Dynamics*, New York: Verlag.)

Over time, three far more efficient forms of written language evolved out of pictographs—Sumerian cuneiform, Egyptian hieroglyphics, and Chinese characters. Egyptian hieroglyphs, which emerged around 2100 BCE, followed a similar development; they used about 700 constructs to produce about 17,000 words. The writing system of the Egyptian language then slowly evolved from hieroglyphs to cursive hieroglyphs and to hieratic, then to demotic forms, and finally to the development and use of Coptic in the first century CE.

Through this transformation, language evolved in several ways. Even as the number of constructs was reduced—Coptic, for example, had a total of 32 constructs, 24 of which were taken from the Greek alphabet—they were able to convey more ideas. Much of this was due to the growing interdependence between written and spoken languages. Instead of representing ideas, the constructs symbolized sounds—like our own alphabet—strung together to reproduce utterances on the page. In addition, the design of constructs became simpler, requiring fewer strokes to draw. The result was that it took less time and energy to convey the message, and the written language became easier to use on a large scale.

230 **This follows from the subsequent argument:** Bejan, A., 2000, *Shape and Structure, from Engineering to Nature*, Cambridge, UK: Cambridge University Press.

232 ***guided locomotion*:** Nature presents us with subterranean and cave-dwelling animals, such as the eastern Mediterranean blind mole rat and the blind cave crayfish, whose evolutionary ancestors were sighted. This is in accordance with the constructal law and our prediction of organ sizes in chapter 3. It predicts that flow systems should morph to enhance their movement for unit of useful energy consumed. Animals that live in environments where it is too dark to see do not need eyes to guide their locomotion.

CHAPTER 10
The Design of History

239 **the biggest flow system that surrounds us: the Earth:** I have no doubt that the design of the cosmos itself—the configuration of stars and planets and the interstices between them—is governed by the constructal law. The reason is that I have applied the constructal law at many scales and in highly diverse domains about which I was curious, and I have found that it works everywhere, in a deterministic sense. The constructal law is valid at all scales for any finite-size flow system that is free to morph. The design of moving mass organized as celestial bodies and interstices should be the same as the design of cracking mud—a solid under volumetric tension that is pulled apart and separates into a conglomerate pattern of mass bodies and interstices (see Bejan and Lorente, 2008, p. 462).

241 **climate zones, temperatures:** Clausse, M., Meunier, F., Reis, A. H., and Bejan, A., 2011, Climate change, in the framework of the Constructal Law, *Earth System Dynamics Discussions*, vol. 2, pp. 241–270; Reis, A. H., and Bejan, A., 2006, Constructal theory of global circulation and climate, *International Journal of Heat and Mass Transfer*, vol. 49, pp. 1857–1875.

241 **does so because it is driven:** Bejan, A., 2006, *Advanced Engineering Thermodynamics*, 3rd ed., Hoboken, NJ: Wiley; Bejan, A., and Lorente, S., 2010, The constructal law of design and evolution in nature, *Philosophical Transactions of the Royal Society B*, vol. 365,

pp. 1335–1347; Bejan, A., and Lorente, S., 2011, The constructal law and the evolution of design in nature, *Physics of Life Reviews,* vol. 8.

242 They are thought of as free: Bejan, A., 2004, *Convection Heat Transfer,* 3rd ed., Hoboken, NJ: Wiley.

243 exergy: Bejan, A., 2006, *Advanced Engineering Thermodyamics,* 3rd ed., Hoboken, NJ: Wiley. The exergy of the Q stream is roughly equal to $Q\,(1 - T_e/T_s) \cong Q$, where T_e and T_s are the sun and Earth temperatures, $T_e << T_s$.

245 always been evolving: Ibid.

246 The conflict between the two camps is real: In biology (animal design evolution) and engineering (human-and-machine species evolution), scientists speak in terms of entropy generation rate *minimization.* In geophysics, scientists speak of entropy generation rate *maximization.* These ad hoc statements of "end design" have limited applicability, and because of this neither is a law: One contradicts the other. The constructal law accounts for both statements. It makes peace between them and corrects them: There is no "min" and "max," and no end design and destiny in nature or in the constructal law. At best, one can speak of evolution toward lower or higher entropy generation rates, provided that one states clearly about which part of Figure 59 one speaks, the engines or the brakes. (Bejan, A., and Lorente, S., 2010, The constructal law of design and evolution in nature, *Philosophical Transactions of the Royal Society B,* vol. 365, pp. 1335–1347; Bejan, A., and Lorente, S., 2011, The constructal law and the evolution of design in nature, *Physics of Life Reviews,* vol. 8; Bejan, A., and Marden, J. H., 2009, The constructal unification of biological and geophysical design, *Physics of Life Reviews,* vol. 6, pp. 85–102.)

250 In 2009, researchers at Princeton and Northwestern: Sokolov, A., Apodaca, M. M., Grzybowski, B. A., and Aranson, I. S., 2010, Swimming bacteria power microscopic gears. *PNAS,* vol. 107, no. 3, pp. 969–974.

254 Big animals and big trucks: Lorente, S., and Bejan, A., 2010, Few large and many small: hierarchy in movement on Earth, *International Journal of Design & Nature and Ecodynamics,* vol. 5, no. 3, pp. 254–267.

259 respective amounts of fuel usage: Bejan, A., and Lorente, S., 2011, The constructal law and the evolution of design in nature, *Physics of Life Reviews,* vol. 8.

SELECT BIBLIOGRAPHY

Amoozegar, Cyrus. 2007. "Constructal Theory of Written Language." In *Constructal Theory of Social Dynamics*, edited by Adrian Bejan and Gilbert W. Merkx. New York: Springer, pp. 297–314.

Bejan, Adrian. 1982. *Entropy Generation Through Heat and Fluid Flow*. New York: Wiley, p. 35.

———. 1996. *Entropy Generation Minimization: The Method of Thermodynamic Optimization of Finite-Size Systems and Finite-Time Processes*. Boca Raton, FL: CRC Press.

———. 1997. "Constructal Theory of Organization in Nature." In *Advanced Engineering Thermodynamics*. 2nd ed. New York: Wiley, pp. 704–811.

———. 2000. *Shape and Structure, from Engineering to Nature*. Cambridge, UK: Cambridge University Press.

———. 2006. *Advanced Engineering Thermodynamics*. 3rd ed. Hoboken, NJ: Wiley, p. 770.

———. 2007. "Why University Rankings Do Not Change: Education as a Natural Hierarchical Flow Architecture." *International Journal of Design & Nature and Ecodynamics* 2 (4): 319–27.

———. 2008. "Constructal Self-Organization of Research: Empire Building Versus the Individual Investigator." *International Journal of Design & Nature and Ecodynamics* 3 (3): 177–89.

———. 2009a. "Two Hierarchies in Science: The Free Flow of Ideas and the Academy." *International Journal of Design & Nature and Ecodynamics* 4 (4): 386–94.

———. 2009b. "The Golden Ratio Predicted: Vision, Cognition and Locomotion as a Single Design in Nature." *International Journal of Design & Nature and Ecodynamics* 4 (2): 97–104.

———. 2010. "The Constructal-Law Origin of the Wheel, Size, and Skeleton in Animal Design." *American Journal of Physics* 78 (7): 692–99.

Bejan, A., and D. Gobin. 2006. "Constructal Theory of Droplet Impact Geometry." *International Journal of Heat and Mass Transfer* 49: 2412–19.

Bejan, A., and P. Haynsworth. 2011. "The Natural Design of Hierarchy: Basketball Versus Academia." *International Journal of Design & Nature and Ecodynamics* 6.

Bejan, A., Edward C. Jones, and Jordan D. Charles. 2010. "The Evolution of Speed in Athletics: Why the Fastest Runners Are Black and Swimmers White." *International Journal of Design & Nature and Ecodynamics* 5 (3): 199–211.

Bejan, A., and S. Lorente. 2001. "Thermodynamic Optimization of Flow Geometry in Mechanical and Civil Engineering." *Journal of Non-Equilibrium Thermodynamics* 26: 305–54.

———. 2004. "The Constructal Law and the Thermodynamics of Flow Systems with Configuration." *International Journal of Heat and Mass Transfer* 47: 3203–14.

———. 2005. *La Loi Constructale.* Paris: L'Harmattan.

———. 2006. "Constructal Theory of Generation of Configuration in Nature and Engineering." *Journal of Applied Physics* 100, article 041301.

———. 2008. *Design with Constructal Theory.* Hoboken, NJ: Wiley.

———. 2010. "The Constructal Law of Design and Evolution in Nature." *Philosophical Transactions of the Royal Society B,* 365: 1335–47.

———. 2011. "The Constructal Law and the Evolution of Design in Nature." *Physics of Life Reviews* 8: 209–40.

Bejan, A., S. Lorente, and J. Lee. 2008. "Unifying Constructal Theory of Tree Roots, Canopies and Forests." *Journal of Theoretical Biology* 254: 529–40.

Bejan, A., S. Lorente, A. F. Miguel, and A. H. Reis. 2006. "Constructal Theory of Distribution of City Sizes." In Bejan 2006, pp. 774–79.

———. 2006. "Constructal Theory of Distribution of River Sizes." In Bejan 2006, pp. 779–82.

Bejan, Adrian, and James H. Marden. 2006. "Unifying Constructal Theory for Scale Effects in Running, Swimming, and Flying." *Journal of Experimental Biology* 209: 238–48.

———. 2009. "The Constructal Unification of Biological and Geophysical Design." *Physics of Life Reviews* 6: 85–102.

Bejan, Adrian, and Gilbert W. Merkx, eds. 2007. *Constructal Theory of Social Dynamics*. New York: Springer.

Bejan, A., and S. Périn. 2006. "Constructal Theory of Egyptian Pyramids and Flow Fossils in General." In Bejan 2006, pp. 782–88.

Biserni, C., L. A. O. Rocha, G. Stanescu, and E. Lorenzini. 2007. "Constructal H-Shaped Cavities According to Bejan's Theory." *International Journal of Heat and Mass Transfer* 50: 2132–38.

Charles, Jordan D., and Adrian Bejan. 2009. "The Evolution of Speed, Size and Shape in Modern Athletics." *Journal of Experimental Biology* 212: 2419–25.

Francis, W. Nelson, and Henry Kučera. 1982. *Frequency Analysis of English Usage: Lexicon and Grammar.* Boston: Houghton Mifflin Company.

Hoppeler, Hans, and Ewald R. Weibel. 2005. "Scaling Functions to Body Size: Theories and Facts." *Journal of Experimental Biology* 208 (9): 1573–74. Special issue.

Lewins, Jeffery. 2003. "Bejan's Constructal Theory of Equal Potential Distribution." *International Journal of Heat and Mass Transfer* 46: 1541–43.

Lorente, Sylvie. 2007. "Tree Flow Networks in Urban Design." In Bejan and Merkx, pp. 51–70.

Lorente, Sylvie, and Adrian Bejan. 2010. "Few Large and Many Small: Hierarchy in Movement on Earth." *International Journal of Design & Nature and Ecodynamics* 5 (3): 254–67.

Lorenzini, Giulio, and Luiz Alberto Oliveira Rocha. 2006. "Constructal Design of Y-Shaped Assembly of Fins." *International Journal of Heat and Mass Transfer* 49: 4552–57.

Manton, Kenneth G., Kenneth C. Land, and Eric Stallard. 2007. "Human Aging and Mortality." In Bejan and Merkx, pp. 183–96.

Miguel, António F. 2006. "Constructal Pattern Formation in Stony Corals, Bacterial Colonies and Plant Roots under Different Hydrodynamics Conditions." *Journal of Theoretical Biology* 242: 954–61.

———. 2010. "Natural Flow Systems: Acquiring Their Constructal Morphology." *International Journal of Design & Nature and Ecodynamics* 5 (3): 230–41.

Morroni, Franca. 2007. "Constructal Approach to Company Sustainability." In Bejan and Merkx, pp. 263–78.

Périn, Stephen. 2007. "The Constructal Nature of the Air Traffic System." In Bejan and Merkx, pp. 119–45.

Poirier, H. 2003. "Une théorie explique l'intelligence de la nature." *Science & Vie* 1034: 44–63.

Quéré, S. 2010. "Constructal Theory of Plate Tectonics." *International Journal of Design & Nature and Ecodynamics* 5 (3): 242–53.

Raja, V. Arun Prasad, Tanmay Basak, and Sarit Kumar Das. 2008. "Thermal Performance of a Multi-block Heat Exchanger Designed on the Basis of Bejan's Constructal Theory." *International Journal of Heat and Mass Transfer* 51: 3582–94.

Reis, A. Heitor, and Adrian Bejan. 2006. "Constructal Theory of Global Circulation and Climate." *International Journal of Heat and Mass Transfer* 49: 1857–75.

Reis, A. Heitor, and Cristina Gama. 2010. "Sand Size Versus Beachface Slope—An Explanation Based on the Constructal Law." *Geomorphology* 114: 276–83.

Reis, A. H., A. F. Miguel, and M. Aydin. 2004. "Constructal Theory of Flow Architecture of the Lungs." *Medical Physics* 31: 1135–40.

Rocha, L. A. O., E. Lorenzini, and C. Biserni. 2005. "Geometric Optimization of Shapes on the Basis of Bejan's Constructal Theory." *International Communications in Heat and Mass Transfer* 32: 1281–88.

Staddon, John E. R. 2007. "Is Animal Learning Optimal?" In Bejan and Merkx, pp. 161–67.

Tiryakian, Edward A. 2007. "Sociological Theory, Constructal Theory, and Globalization." In Bejan and Merkx, pp. 147–60.

Weibel, Ewald R. 2000. *Symmorphosis: On Form and Function in Shaping Life*. Cambridge, MA: Harvard University Press.

Weinerth, G. 2010. "The Constructal Analysis of Warfare." *International Journal of Design & Nature and Ecodynamics* 5 (3): 268–76.

www.constructal.org.

www.isihighlycited.com.

www.mems.duke.edu/fds/pratt/MEMS/faculty/abejan.

Zipf, G. K. 1949. *Human Behavior and the Principle of Least Effort: An Introduction to Human Ecology*. Cambridge, MA: Addison-Wesley.

INDEX

Page numbers in *italics* refer to illustrations.

ABOUT THE AUTHORS

ADRIAN BEJAN has pioneered numerous original methods in thermodynamics, such as entropy generation minimization, scale analysis of convection, heatlines and masslines, and the constructal law of design and evolution in nature. He is ranked among the hundred most cited authors in all engineering (all fields, all countries, living and deceased) by the Institute of Scientific Information (isihighlycited.com). His h-index is 45.

He is the author of more than 540 peer-reviewed journal articles and 24 books, including *Shape and Structure, from Engineering to Nature* (Cambridge University Press, 2000), *Constructal Theory of Social Dynamics* (Springer, 2007), and *Design with Constructal Theory* (Wiley, 2008). His treatises *Advanced Engineering Thermodynamics* (Wiley) and *Convection Heat Transfer* (Wiley) are now in their third editions and are used as graduate textbooks in universities around the world.

He has been awarded sixteen honorary doctorates by universities in eleven countries, including the Swiss Federal Institute of Technology (2003) and the Sapienza University of Rome (2009). He has received numerous national and international society awards, such as:

Max Jakob Memorial Award (American Society of Mechanical Engineers & American Institute of Chemical Engineers, 1999)

Donald Q. Kern Award (American Institute of Chemical Engineers, 2008)
Honorary Member, American Society of Mechanical Engineers, 2011
Ralph Coats Roe Award (American Society for Engineering Education, 2000)
Luikov Medal (International Centre for Heat and Mass Transfer, 2006)
James P. Hartnett Memorial Award (International Centre for Heat and Mass Transfer, 2007)
Edward F. Obert Award (American Society of Mechanical Engineers, 2004)
Worcester Reed Warner Medal (American Society of Mechanical Engineers, 1996)
James Harry Potter Gold Medal (American Society of Mechanical Engineers, 1990)
Charles Russ Richards Memorial Award (American Society of Mechanical Engineers, 2001)
Gustus L. Larson Memorial Award (American Society of Mechanical Engineers, 1988)

Bejan received all his degrees from MIT: BS (1971, Honors Course), MS (1972, Honors Course), and PhD (1975). He was a postdoctoral fellow at the University of California, Berkeley, at the Miller Institute of Basic Research in Science (1976–1978). He was appointed as a full professor of mechanical engineering with tenure at Duke University in 1984, and J. A. Jones Distinguished Professor in 1989.

J. PEDER ZANE is an assistant professor of journalism and mass communications at St. Augustine's College in Raleigh, North Carolina. He is an award-winning columnist who has worked for the *New York Times* and the *News & Observer* (Raleigh). He has edited and contributed to two books for W. W. Norton, *Remarkable Reads: 34 Writers and Their Adventures in Reading* (2004) and *The Top Ten: Writers Pick Their Favorite Books* (2007).